CLIVE V.J. WELHAM

The Evolution of Asexual Reproduction in Plants

Michael Mogie

School of Biological Sciences, University of Bath, UK

CHAPMAN & HALL

London · Glasgow · New York · Tokyo · Melbourne · Madras

Published by Chapman & Hall, 2–6 Boundary Row, London SE1 8HN

Chapman & Hall, 2–6 Boundary Row, London SE1 8HN, UK

Chapman & Hall, 29 West 35th Street, New York NY10001, USA

Chapman & Hall Japan, Thomson Publishing Japan, Hirakawacho Nemoto Building, 6F, 1–7–11 Hirakawa-cho, Chiyoda-ku, Tokyo 102, Japan

Chapman & Hall Australia, Thomas Nelson Australia, 102 Dodds Street, South Melbourne, Victoria 3205, Australia

Chapman & Hall India, R. Seshadri, 32 Second Main Road, CIT East, Madras 600 035, India

First edition 1992

© 1992 M. Mogie

Typeset in 10/12pt Trump Mediaeval by Intype, London
Printed in Great Britain by St Edmundsbury Press, Bury St Edmunds, Suffolk

ISBN 0 412 44220 5

A catalogue record for this book is available from the British Library

Library of Congress Cataloging-in-Publication data
Mogie, Michael.
    The evolution of asexual reproduction in plants / by Michael Mogie.
        p.   cm.
    Includes bibliographical references and index.
    ISBN 0–412–44220–5 (hard)
    1. Plants—Reproduction. 2. Plants—Evolution.
3. Reproduction, Asexual. I. Title.
QK826.M64   1992
581.1'65—dc20   91–41024
                                                                CIP

To Janis, Richard and James

Contents

Acknowledgements

I wish to thank the *Biological Journal of the Linnean Society* for permission to reproduce, in whole or in part, Tables 5.1, 5.3, 6.2, 6.5, 6.6 and 6.7, which have appeared in Mogie (1985, 1988) and Mogie and Ford (1988).

Preface

Biology is encumbered with terms that are widely used but which are given different meanings by different users. 'Asexual reproduction' is an example. Because of this, it is important that, at the outset, I state the definition of asexual reproduction that will be used in this book, and attach an apology to it, for I am in no doubt that some readers will see my definition as too narrow.

The organisms around which this book is based are multicellular and embryo-producing. I define asexual reproduction in these as comprising the production of an embryo from a single cell whose nucleus is not formed by syngamy.

By stipulating that the initial product of asexual reproduction must be an embryo I am *de facto* excluding vegetative propagation (by stolons, rhizomes, bulbs etc.) as a process of asexual reproduction. I defend my position on this issue in the first chapter, but a good working defence that is worth pointing out now is that, from an evolutionary and ecological perspective, the asexual production of embryos is a very different phenomenon to the replication of a genotype through vegetative propagation. The former is overtly a reproductive process that offers a direct challenge to sexual reproduction, and need not entail the accurate replication of the maternal genome (for example, in the case of haploid parthenogenesis in the social insects). It is with this direct challenge to embryo production by sexual reproduction that most of the debate in recent years on the relative merits of sexual and asexual reproduction has been concerned. And it is with this direct challenge that a sizeable proportion of this book is concerned. In contrast, vegetative propagation is, in my view, a growth process that offers, at best, an indirect challenge to sexual reproduction – demonstrated by the fact that most plants that undertake vegetative propagation produce embryos sexually.

Within the confines of this definition of asexual reproduction it can be shown that, among land plants, asexual reproduction is exhibited by many flowering plants and ferns but is effectively absent from bryophytes and gymnosperms. But even within the ferns and flowering plants asexual reproduction is much less common than sexual reproduction. Moreover, most asexual flowering plants initiate embryos from unfertilized eggs, but embryos produced by asexual ferns are initiated from somatic cells. And whereas asexual reproduction in ferns is usually obligate, many asexual flowering plants are facultatively asexual, with each individual producing a proportion of its progeny asexually from meiotically unreduced eggs or from somatic cells, and the remainder sexually from fertilized, meiotically reduced eggs.

Clearly, this variation must be rooted in the different biologies of these groups. By identifying the different aspects of these biologies that are responsible, important insights can be gained into the causes and consequences of sexual and asexual reproduction. I have attempted to identify these in this book. It is a task that has given me considerable pleasure, leading me on invigorating journeys through phylogeny, ontogeny, ecology and genetics. I hope it has been one of useful discovery.

Much of the recent debate about sexual and asexual reproduction has centred on the costs and benefits of the latter compared with the former. I have fully acknowledged this approach. But I have not been unduly constrained by it. Indeed, I argue that other issues must often be given prominence if a thorough understanding is to be obtained of the factors influencing the evolution, establishment and maintenance of asexual reproduction. Thus I question the view that asexual reproduction is necessarily primarily selected as an alternative to sexual reproduction in anisogamous organisms, and I propose that the incidence and taxonomic distribution of asexual reproduction *per se* and of its different forms can often be most clearly understood in terms of phylogenetic and ontogenetic constraints rather than in terms of cost-benefit analysis.

It goes almost without saying that asexuality is fundamentally a phenomenon that affects the female function. But this should not mean that we can ignore the fact that most asexual plants are descended from cosexual sexual ancestors and that many continue to produce some viable male gametes. Indeed, the important role the male function can play in the spread of a gene for asexual reproduction is recognized. But it has gradually become clear to me that its role goes much deeper and wider than this, enabling genes for asexuality to seek out the best genomes, providing established asexual populations with a first and highly effective line of defence against incursion by related sexual forms, and helping to shape the geographic distribution of related sexual and asexual forms. Indeed, the role of the male function in the biology of asexuality in cosexual forms can be so important that I have felt that it deserves special prominence. Consequently, I have devoted an entire chapter (Chapter 4) to it.

Of course, I could be mistaken in many of my arguments, and I cheerfully admit that some of the hypotheses I propose rest on rather insubstantial foundations. But a large part of the enjoyment of this investigation has been to enter areas where few have ventured before, and I offer my views in the belief that their basic framework, at least, is sound.

During the planning of this book I came to the decision that I would keep extraneous detail to a minimum – that I would not offer a comprehensive review either of asexual reproduction in general or of how it relates specifically to plants. Instead, I would review only those aspects of asexuality that were central to the development of my arguments on the evolution, establishment and maintenance of asexual reproduction. Consequently, there are several aspects of asexuality about which I give little detail. I steer well clear of the difficulties that envelop the taxonomic description of asexual organisms and I deal only with those aspects of cytology and of the development of the female gametophyte that impinge directly on the lines of argument that I develop. Sloth has no doubt played some part in my decision to restrict the breadth of the book, as has a lack of confidence that I could contribute anything useful to certain areas. But my main reasons for adopting this approach are that I did not wish to divert myself, or risk unnecessarily diverting the reader, from the main issues being addressed and, more prosaically, I felt rescued from the need to provide a comprehensive coverage of all aspects of asexual reproduction in plants by the publication over the last few years of a number of excellent and easily accessible reviews that offer timely discussions of many of these areas and issues.

The arguments presented here have emerged gradually over the past few years, and I am indebted to many colleagues who have offered support and advice during this period. I would especially like to thank John (A.J.) Richards, of the University of Newcastle, who first aroused and then nurtured my interest in asexual reproduction by the excellent and expert undergraduate teaching and postgraduate supervision he provided, John Maynard Smith, of the University of Sussex, in whose stimulating company I spent an immensely enjoyable period undertaking postdoctoral research, and Nigel Franks, Alan Rayner, Keith Moore and Henry Ford of the University of Bath. I have also been very fortunate in entering into correspondence with G. Ledyard Stebbins of the University of California on the geographic distribution of apomicts and with Yves Savidan of ORSTOM on the genetic control of apomixis, and in receiving the views of Richard Abbott and Peter Gibbs of the University of St Andrews on the arguments I present on growth and reproduction in the first part of this book. However, the good advice that has been offered to me may not always have been taken, and any inconsistencies or mistakes are my own.

Michael Mogie
University of Bath

Chapter 1

Prologue

1.1 PREAMBLE

It is the primary function of every organism to pass on to the next generation as many copies of its genes as possible. This function can be fulfilled through an organism's own reproductive effort and, as Hamilton (1964a,b) points out, through the organism acting to enhance the reproductive success of relatives, with which it shares genes in common. The inevitability of death guarantees the status of this function. A gene that contributes to the reproductive success of its carriers helps to ensure that it survives their demise. Those that are more successful in doing so will survive in greater numbers, and will eventually replace those that are less successful. In this way, genes which cause their carriers to maximize their representation in the gene pool of the next generation will come to predominate.

A major consequence of this scramble for reproductive success has been the evolution of a multitude of ecological, behavioural and reproductive strategies. Some affect the reproductive process directly, honing the organism into an efficient seeker and attractor of mates, or into an efficient producer of offspring. Others are less direct, providing the organism with the capacity to grow to meet, or to survive to fulfil, its reproductive potential. All combine to produce a plexus of life cycles and life history strategies – an ocean of solutions to endless streams of problems, providing evolutionary and population biologists with a complex but invigorating habitat in which to wade, swim and sometimes sink without trace!

Among the many strategies that have emerged are those that manipulate the process of initiating progeny. In the majority of eukaryotic organisms, an individual is initiated from a cell whose nucleus is the product of the fusion of two nuclei. This pattern of reproduction is usually described as sexual. However, in a minority of taxa, the initiation of progeny is not dependent on the fusion of nuclei. Instead, the mother differentiates a cell which acts directly as the initial cell of the offspring. This pattern of reproduction is usually – though, as we shall see, by no means consistently – described as asexual. It is clear that asexually reproducing eukaryotes are descended from sexual ancestors (Gustafsson, 1946–47). There is much morphological, anatomical and developmental evidence demonstrating this (e.g. Richards, 1973). For example, autonomously apomictic members of flowering plant genera whose sexual members rely on insect pollination still produce structures and substances associated with pollinator attraction (e.g. petals and nectar) even though they do not need to attract pollinators to achieve reproductive success. But although it is clear that sexuality can give way to asexuality, what is not clear, or is less clear, is why, when or how a transition to asexuality occurs. The first two parts of this problem have been intensively investigated, and proposed solutions have been keenly debated. The question of how it arises has received much less attention. And yet, an appreciation of this aspect of the problem is essential to a thorough understanding of the issue. In the pages that follow, I hope to demonstrate this, by investigating events associated with the emergence of asexuality in plants and by showing how the knowledge gained can aid understanding of other aspects of asexuality.

It is perhaps pertinent at this stage to describe how the book is structured, and to introduce the issues that will be covered. The plant groups that will be investigated are the embryophytes: the bryophytes and tracheophytes. These produce embryos within multicellular reproductive organs; the embryos are dependent, at least initially, on maternal tissues for protection and nutritional support. Attention will be focused within the bryophytes on the mosses and liverworts and within the tracheophytes on the ferns, gymnosperms and angiosperms (flowering plants).

This chapter will conclude with a general discussion about reproduction and its sexual and asexual forms. This is a necessary diversion from the main subject matter of the book for, as noted above, although the terms are used widely, they are not applied consistently. The main purposes of this discussion are to illustrate how reproduction and its sexual and asexual forms will be defined, and to provide the reasons for this choice. The definitions offered, and the arguments used to defend them, may or may not be widely acceptable, but they will serve their purpose as long as they are clear.

Two interrelated themes will then be developed. The first centres on the issue of when asexuality should emerge as an adaptive strategy, and with the types of asexual reproduction that are most likely to become established in different taxa. The second revolves around the problems that must be overcome if a form of asexual reproduction is to evolve. These themes share ground in common, but it is nevertheless worthwhile to approach them separately. This part of the discussion will be approached via Chapter 2, which provides an account of sexual and asexual reproduction in embryophytes. A comprehensive review is not attempted. Instead, effort is directed towards highlighting only those aspects of reproduction that are relevant to the arguments developed in subsequent chapters. The anomalies in the incidence of asexuality, and the taxonomic clustering of types of asexual reproduction, that will be exposed in this chapter will be explained in Chapter 3.

Most asexual plants are descended from cosexual ancestors (i.e. each individual produced both male and female gametes) and still exhibit a male function. Perhaps counter-intuitively, the male function in these has played, and continues to play, an important, sometimes central, role in the establishment and maintenance of asexual reproduction. I will argue that it has also greatly influenced the quality of the genetic background against which genes for apomixis must operate, that it has influenced the geographic distribution of related sexual and asexual forms and that in many taxa it continues to have an important effect on the viability and sometimes on the vigour of asexually reproduced offspring. The importance of the male function is under-appreciated. Amends will be made in this book by devoting the whole of Chapter 4 to it.

The second theme to be developed, which involves an investigation of the problems that must be overcome if a form of asexual reproduction is to evolve, is the subject matter of Chapter 5. Unfortunately, most of the data relevant to this issue have emerged from studies of asexuality in flowering plants. It is necessary to reflect this taxonomic bias. This chapter will consequently concentrate on examining the problems faced by this group in evolving parthenogenetic forms of generative (diplosporous) and aposporous apomixis – the forms of asexual reproduction that are most common in this group.

The ultimate message of Chapters 3–5 is that the conditions favouring the evolution of asexuality, and the abilities of organisms to evolve different forms of asexual reproduction, are more restricted than is generally appreciated. Two further issues will be considered in Chapter 6. The first is one that has seen biologists of various disciplines locked into an intellectual scrummage. It concerns the maintenance of genetic recombination – an issue that is considered to be almost synonymous with that of the maintenance of sex. I will argue that current explanations of the maintenance of genetic recombination are most valid when obligately sexual and obligately asexual taxa are being compared and that they fail to explain why genetic recombination should be maintained by facultative apomicts during sexual reproduction but not during asexual reproduction. An alternative explanation for the maintenance of genetic recombination will be tentatively proffered.

The second issue discussed in Chapter 6 concerns the capacity of asexual organisms to generate mutations and the consequences of this in terms of their evolutionary potential. The arguments put forward in this and other chapters will be brought together and subjected to some further analysis in Chapter 7.

Although the overt aim of this book is to offer an interpretation of asexual reproduction in plants, many of the arguments presented will be bolstered with reference to studies involving other taxa. On several occasions, these studies will take centre stage. This approach is absolutely necessary. Insufficient is known about asexuality to allow an understanding of its control to be approached from information gleaned from a single kingdom. However, if, in pursuit of this information, the sleuth is willing to prowl other kingdoms, solutions will emerge to problems that would have remained as such had the cosmopolitan approach not been followed. A danger of this approach is that comparisons between widely separated taxa can be spurious, and hypotheses based on them consequently weakened or invalidated. However, I believe that I have exercised sufficient caution, and sampled adequately, during my collecting trips for this not to be a problem.

1.2　REPRODUCTION, GROWTH AND THE INDIVIDUAL

It is an interesting and informative exercise to hunt through widely read texts tracking definitions of sexual and asexual reproduction. It soon becomes clear that there is considerable, and sometimes fundamental, disagreement over the essence of these phenomena. Some definitions are obviously awry through taxonomic bias, describing sexual reproduction as though it is a phenomenon confined to diploid, anisogamous, and often obligately outcrossing, organisms. These definitions are obviously not meant to be exhaustive (although this is often not stated) but simply to be relevant to a particular taxon (although this is often not made clear), and it seems unfair to highlight any single example. Hopefully, their limitations are apparent to their readers.

Such definitions are often to be found in texts describing basic principles of evolutionary or population biology. However, major differences of opinion also separate texts exclusively devoted to investigations of the significance of sexual and asexual reproduction (e.g. Williams, 1975; Bell, 1982). These differences affect the status of several forms of reproduction. Given this situation, it would be unwise simply to proceed with a discussion of sexual and asexual reproduction as though their definitions were generally agreed. To do so would surely lead to confusion, frustration and irritation. Thus the major purpose of the remainder of this chapter is to try to formulate reasonable and usable definitions of these processes. Unfortunately, it is no easy matter even to begin this task for, when launching into any debate about the essential nature of biological processes it is typical to find the slipway receding into other areas of controversy. In this case, the prior issue that must be resolved before the issue of which forms of reproduction are sexual and which asexual concerns the definition of reproduction *per se*. What do we mean by reproduction, and how does it differ from growth?

Most definitions of reproduction can be reduced to the statement that it involves the generation of a new individual. However, biologists cannot agree a definition of an individual! We cannot even agree about whether, *de facto*, the product of reproduction must be a free-living organism, becoming physiologically and physically independent of its mother, either immediately or after a period of nurture.

The flavour of this part of the debate can be savoured from a sample of some recent and not so recent publications. We can begin with Huxley (1912) who viewed the individual as a unit that would be non-functional if it was cut in half. This 'indivisibility criterion' (Dawkins, 1982) is one that concurs with what may be described as our emotional view of individuality in higher animals. Thus, what develops by mitosis from the zygote in a mammal is a physiologically integrated and physically distinct unitary body that exhibits determinate growth. A mammal grows by increasing the size of a genetically predetermined number of parts (legs, eyes, lungs, ovaries etc.), not by accumulating extra copies of parts. Because of this, it will not remain viable if it is subdivided. It meets the indivisibility criterion. But this criterion fits other groups far less well. For example, what develops from the zygote in plants, and in some animals (e.g. hydrozoans) is a modular structure of indeterminate growth (a module is the 'multicellular unit of construction that is iterated in the process of growth' (Harper *et al.*, 1986). Thus growth in a plant does not comprise an increase in size of a genetically predetermined number of parts. Rather, it includes the accumulation of extra copies of parts (extra leaves, buds, stems, ovules etc.). In many plants modules can persist as physically and physiologically independent units. This capacity is frequently realized in clonal/vegetative propagation. Thus what develops from the zygote in modular organisms does not meet the indivisibility criterion. Under this criterion, a plant zygote can give rise to a host of individuals.

In stark contrast to Huxley, Harper (1977:27) offers a definition of growth, reproduction and individuality that acknowledges the modular nature of many organisms. He argues that, for plants,

The distinction made here between 'reproduction' and 'growth' is that reproduction involves the formation of a new individual from a single cell: this is usually (though not always e.g. apomicts) a zygote. In this process a new individual is 'reproduced' by the information that is coded in that cell. Growth, in contrast, results from the development of organized meristems. Clones are formed by growth – not reproduction.

Using Harper's terminology, the individual is a 'genet' which, in modular organisms comprises all of the modules descended from the same zygote. In clonal organisms, the 'ramet' is the unit of clonal growth, the module that is capable of following an independent existence if separated from the rest of the genet. Although this terminology has been widely adopted, the view that ramets are grown, not reproduced, has found support only among some biologists. Recent publications which have described clonal propagation as growth include Dawkins (1982), Tiffney and Niklas (1985) and the contributing authors to the volume on clonal propagation edited by Van Groenendael and de Kroon (1990). These can be contrasted with Grime (1979), Bell (1982), Mishler (1988) and Maynard Smith (1989) who describe it as reproduction. Schmid (1990) makes no absolute distinction between the two processes, describing modular organisms as being able to '. . . grow clonally by vegetative reproduction of modular structures'. However, he does make a distinction between vegetative and asexual reproduction, defining the former as clonal growth and the latter (which includes apomixis and parthenogenesis) as '. . . a modification of sexual reproduction . . .'.

Harper's view that a single cell is the fundamental unit of reproduction has received a mixed reception in the biological literature. Janzen (1977) concurs with the basic principle but goes much further than Harper in arguing that only one type of single-celled propagule – the zygote – is the unit of reproduction. Other single-celled propagules contribute to growth. Applying the concept of the 'evolutionary individual', he describes a new individual as resulting from the fusion of gametes. During its life cycle, this individual may propagate itself in various ways, some of which may involve non-zygotic single-celled structures, but in so doing it is undergoing growth not reproduction, even if the propagule is a parthenogenetically developing egg. Thus he views an apomictic dandelion (*Taraxacum*) clone as a

   . . . very long-lived perennial organism. At any time, it is composed of parts that are moving around ('seeds' produced by apomixis), growing (juvenile plants), dividing into new parts (flowering plants), and dying (all ages and morphs).

Such disagreement over the nature of growth, reproduction and the individual has led Dawkins (1982:253) to reflect that the organism may be '. . . a concept of dubious utility'. However, in a more optimistic mood he goes on to suggest that a useful, though lateral, way of approaching the problems under discussion is first to identify a consequence of reproduction. In doing so, the difference between the process of reproduction and that of growth will emerge, and the limits of individuality will be circumscribed. Thus Dawkins considers that, uniquely, reproduction provides an opportunity for fundamentally changing the complex structure of organisms. In contrast, growth merely provides the opportunity for minor adjustments to this structure. For reproduction to be able to fulfil this role the fundamental unit of reproduction must be a single cell that gives rise to a lineage of cells that are able to persist in evolutionary time (e.g. germ-line cells). The central theme of his argument is that the single-celled unit qualifies as a reproductive unit because what develops from it by cell division is initiated at the beginning of ontogeny. It is here, when organs are being formed, that novel genetic mutations can have most influence on the pattern of development. Thus Dawkins writes (p. 262)

> . . . the significance of the difference between growth and reproduction is that reproduction permits a new beginning, a new developmental cycle and a new organism which may be an improvement, in terms of the fundamental organization of complex structure, over its predecessor. Of course it may *not* be an improvement, in which case its genetic basis will be eliminated by natural selection. But growth without reproduction does not even allow the *possibility* of radical change at the organ level, either in the direction of improvement or the reverse. It allows only superficial tinkering . . . The point about recurring reproduction life cycles, and hence, by implication, the point about organisms, is that they allow repeated returns to the drawing board during evolutionary time.

Dawkin's argument is an attractive one when it is applied to unitary, determinate organisms but it is less so when it is applied to indeterminate modular organisms. The iterative modular pattern of growth results in the repetition of developmental processes. Moreover, the apical meristems from which modules arise are single-celled in bryophytes and most ferns (Mishler, 1988; Barlow, 1989; Schmid, 1990) and will be able to initiate the type of radical change *within* modular organisms that single-celled reproductive units can initiate *between* unitary organisms. Indeed, Hardwick (1986) has argued that the multicellular apical meristems typical of higher plants may have evolved from single-celled meristems to reduce the potential for change that the latter offer selfish cell lines. Clearly modular growth in organisms with single-celled apical meristems blurs the distinction between reproduction and growth made by Harper and Dawkins. Indeed, this has led Mishler (1988) to question the applicability of Harper's definition to bryophytes, and he argues instead that, in this group, reproduction should be defined simply as the production of a new, physiologically independent plant, irrespective of whether this unit arises from clonal propagation or from the germination, following dispersal, of spores. In contrast, growth is defined simply as an increase in size of a single physiological unit.

The prevalence of single-celled apical meristems in some groups imposes limits on the usefulness of definitions that highlight the role of single-celled propagules in reproduction. But there is a risk that such definitions may be compromised even further in that the life cycle characteristic of bryophytes and tracheophytes includes two single-celled stages: that of the zygote (or parthenogenetically dividing egg) and that of the meiospore. Very briefly, the zygote gives rise by mitosis to a multicellular sporophyte, cells of which undergo meiosis to produce meiospores. Each meiospore gives rise by mitosis to a multicellular gametophyte which differentiates gametes by mitosis. These fuse in pairs to produce zygotes. This life cycle is described as incorporating an alternation of generations, but the literature remains ambiguous about whether this descriptive term is to be interpreted as implying that the sporophyte and the gametophytes produced from it are different individuals or are merely different parts of the same individual.

If we stick rigorously to the view that what emerges from a single cell is reproduced, not grown, then sporophytes and gametophytes must be identified as different individuals. One important consequence of doing this is that it becomes necessary to acknowledge that obligate sexuality is a condition foreign to bryophytes and tracheophytes! We must view most plant taxa as being cyclically sexual/asexual, alternating sexual (gametophytic) and asexual (sporophytic) generations, and the remainder as being obligately asexual (e.g. apomicts, where the gametophyte as well as the sporophyte reproduces asexually). But perhaps sticking rigorously to the view that what emerges from a single cell is reproduced rather than grown is unfair. Implicit in Harper's (1977) and Dawkins' (1982) definitions of reproduction and growth is the proviso that in order to be reproduced, not grown, the unit developing from a single-celled propagule must be capable of surviving as a physiologically and physically discrete unit. I will describe this as the 'independence criterion'. The difficulty here is that the addition of this proviso further clouds, rather than clarifies, the issue of whether sporophytes and gametophytes are to be treated as different individuals.

The extent to which sporophytes and gametophytes meet the independence criterion is dependent on taxonomic location and gender. For example, a characteristic of bryophytes and tracheophytes is that the embryo (progeny sporophyte) is not dispersed away from the maternal gametophyte. Nevertheless (and with the exception of the fern *Anogramma leptophylla* – Mehra and Sanhu, 1976), the tracheophyte sporophyte eventually achieves independence by growing through the gametophyte (which is often ephemeral when compared with the sporophyte) and establishing its own rooting system and photosynthetic apparatus. But the bryophyte sporophyte (which is often ephemeral when compared with the gametophyte) remains attached to, and dependent on, the gametophyte for the whole of its existence. Thus the independence criterion is met by the tracheophyte sporophyte but not by the bryophyte sporophyte. If it is to be applied then it must be concluded that the sporophyte is reproduced by most tracheophytes but is grown by bryophytes. Applying the independence criterion to the embryophyte gametophyte is even more problematic than applying it to the sporophyte. The gametophyte exists independently of the maternal sporophyte whenever meiospores or juvenile gametophytes are dispersed. This is a characteristic of the bryophytes, homosporous ferns and most heterosporous ferns (*Selaginella rupestris* is an exception, as the megagametophyte – the female gametophyte – is retained within the dehisced megasporangium (Sporne, 1975)). It is also characteristic of the male gametophytes (pollen) of gymnosperms and angiosperms (although it is debatable whether the pollen of self-fertilizing plants can be described as having been dispersed from, and having achieved true independence of, the parental sporophyte). However, the female gametophyte of gymnosperms and angiosperms does not become physically and physiologically independent of the maternal sporophyte. It undergoes dispersal, but it does so within a mass of enclosing maternal sporophytic tissue. That is, the dispersal unit (the seed) comprises maternal sporophytic tissue enclosing the female gametophyte within which the progeny sporophyte is developing. (Following dispersal, the progeny sporophyte breaks free from the rest of the seed during germination and achieves a free-living, independent status.) Thus, using the independence criterion, it would have to be argued that the female gametophyte of seed plants is grown by the sporophyte, the male gametophyte is reproduced by the sporophyte (at least if outcrossing occurs), and the sporophyte is reproduced by the female gametophyte! Clearly, applying the independence criterion to the products of single-celled propagules is not a particularly helpful way of distinguishing between reproduction and growth in plants.

Even following such a brief and incomplete review of the problems involved in defining the individual, growth and reproduction, it should be clear that the differences of opinion between biologists about these processes are considerable. Almost any definition offered will satisfy some but will be met with considerable resistance by others. Nevertheless, the subject matter of this book makes it necessary that I declare an opinion on this issue. I will first offer my definitions and then attempt to justify them.

The definitions that I favour are as follows. Reproduction is the event which initiates a recapitulation of ontogeny. That is, reproduction re-sets ontogeny. Growth is part of the process (along with differentiation and development) which causes ontogeny to progress from its earliest to its latest stages. The individual is the unit associated with a single ontogenetic progression.

The definition of reproduction offered, requires that the unit of reproduction is a single cell that, in multicellular organisms, divides to produce an embryo. Cells that can perform this role include zygotes, parthenogenetically developing eggs and apogamously developing gametophyte cells (parthenogenesis and apogamy are described in Chapter 2). Cells that are not zygotes but which can initiate embryos will be described as zygote equivalents. It follows that any single-celled unit of propagation whose product is at an advanced (i.e. postzygotic) stage of ontogeny is contributing to, and is consequently an aspect of, the growth of the individual. This is the case irrespective of whether or not the product is free living. Thus gametophytes are grown by sporophytes, and ramets are grown by genets, because both are advanced stages of ontogeny. It also follows that neither the indivisibility nor the independence criterion applies to the definition of the individual. With respect to the latter criterion, the bryophyte zygote is at the beginning of ontogeny in the same way as the zygote is at the beginning of ontogeny in mammals, ferns or seed plants. Although the sporophyte that develops from the bryophyte zygote remains dependent on the gametophyte, the zygote is nevertheless reproduced, not grown.

In placing ontogeny at the centre of my definitions I am suggesting that a useful way to view growth, reproduction and the individual is from the view-point of a gene. The individual is the carrier of the gene. A gene will survive the death of its carrier only if it is part of a genome that can direct the carrier from the beginning to the end of ontogeny, as only at or towards the end of ontogeny is reproductive maturity achieved. From a gene's point of view, the carrier is the unit that protects and carries it and which is likely to utilize it at some stage of the carrier's ontogeny. If new genomes are to be formed from the population gene pool then they are formed at the beginning of ontogeny. This is also the stage in the life cycle where the efficiency of the genome is first tested. Testing will continue throughout the ontogenetic progression, as different genes will be expressed at different stages in the progression and in different tissues. If the genome survives each test it will graduate at the end of the ontogenetic progression, the genes which comprise it passing into cells – zygotes or zygote equivalents – that are at the earliest stage of another ontogenetic progression. Events like vegetative propagation and gametophyte production that occur at stages along the ontogenetic progression simply put to the test the genome that was brought together at the beginning of ontogeny. Vegetative propagation tests whether the genome can function efficiently through space and/or time. If the genome fails this test the carrier dies before it completes its ontogeny, and the genes that formed the genome will fail to survive the carrier's death. Gametophyte production in sexual species tests whether the genome can survive fragmentation. Once again, if the genome fails the test, the genes that comprise it will fail to survive the carrier's death. Thus from a gene's-eye-view the individual is the carrier, the unit that the genome, of which the gene is part, has to lead through an ontogenetic progression. If a gene is to persist in evolutionary time it will have to survive a series of such progressions. It is therefore useful to draw a line between ontogenetic progressions, defining the unit that undergoes a single progression as an individual, defining growth as part of a single progression, and defining the event that marks the culmination of a progression, the generation of new ontogenetic progressions, as reproduction.

I am in no doubt that my views on reproduction, growth and the individual will fail to satisfy many biologists. Of particular concern may be my decision to include vegetative propagation as a growth process rather than as a form of asexual reproduction. I described earlier in this section how biologists are deeply divided over this issue. By including it as a growth process, I am signalling that I will not be including vegetative propagation in the discussions that follow on the evolution, maintenance, costs and benefits of asexual reproduction in plants. But I would argue that even if the reader views vegetative propagation as asexual reproduction its omission should be seen more as an irritation than as an error. A fundamental distinction should be maintained between vegetative propagation by stolons, bulbils, fragmentation etc. and the asexual production of embryos. Only the latter of the two processes offers a direct challenge to sexual reproduction; most vegetatively propagating plants produce their embryos by sexual reproduction. Certainly, the major debates in the literature

about the advantages and disadvantages of sexual and asexual reproduction have centred on the issue of the advantages and disadvantages
of producing embryos by sexual and asexual processes (e.g. Maynard
Smith, 1978; Bell, 1982; Michod and Levin, 1988). Vegetative propagation does not offer the challenge to sexual reproduction that is
offered by apomixis in flowering plants or agamospermy in ferns. It is
with this challenge that this book is primarily concerned.

1.3   SEXUAL AND ASEXUAL REPRODUCTION

Having spent some time discussing the difference between reproduction and growth it is now necessary to do the same for sexual and
asexual reproduction, although in doing so I will not be returning to
the debate over whether vegetative propagation is a form of asexual
reproduction or of growth. The need for this discussion will be made
evident by referring to the views of Williams (1975) and Bell (1982)
who, in influential and important texts, provide very different definitions of sexual and asexual processes. Following this comparison,
my own approach to this problem will be described.

Williams (1975: 114–117) describes a scheme in which the distinction between sexual and asexual reproduction is made on the grounds
of whether reproduction generates genetically novel offspring. Three
categories of sexual reproduction and two of asexual reproduction
are recognized. The categories of sexual reproduction are: primordial,
euphrasic and degenerate. Primordial sexual reproduction includes
transduction and transformation in bacteria. It results whenever mechanisms of genotypic maintenance are poorly developed and lead to
genetically variable progeny. Here, sexual reproduction appears to be
an unavoidable consequence of existence, in much the same way
as ageing is an unavoidable consequence of survival, rather than a
phenomenon to be acquired or discarded. Even if it was advantageous
for these organisms to reproduce by simple fission, their inability
adequately to maintain their genotypes makes sexual reproduction
inescapable. Euphrasic sexual reproduction is defined as requiring the

> . . . production of haploid gametes that must unite in pairs to
> form an at least momentarily diploid zygote. The gametes that
> unite must come from different individuals so that the zygote
> generation is genetically diverse . . .

It may be conservative or costly. The conservative type includes those
forms of reproduction that involve no loss of genetic material during
meiosis. That is, each meiosis gives rise to four gametes, with each
possessing a different member of the four nuclei of the meiotic tetrad.

Williams lists euphrasic reproduction in isogamous organisms as conservative, but so too is euphrasic reproduction in (anisogamous) bryophytes and homosporous ferns, as these produce four spores per meiosis each of which can give rise mitotically to gamete-producing gametophytes. The costly type of euphrasic reproduction includes those forms of reproduction that involve a 50% loss of genetic material during female meiosis. It is practised by anisogamous animals and most heterosporous plants, where only one of the four tetrad nuclei develops into the egg nucleus, the other three either degenerating or otherwise being excluded from the egg. The fern genus *Selaginella* is an exception among heterosporous plants, as more than one functional megaspore may be produced by the female meiosis (Sporne, 1975). Degenerate sexual reproduction includes selfing by hermaphrodites and '. . . parthenogenesis, whenever the process fails to preserve the maternal genotype . . .' including the '. . . artificially induced development of haploid eggs . . .'

By extension, it will also include haploid parthenogenesis in arrhenotokous taxa, as long as the organism producing the egg is allozygous at one locus at least. (Arrhenotoky is the phenomenon whereby the females of a species are produced from fertilized eggs and are consequently diploid, and the males are produced from parthenogenetically developing haploid eggs. It is found within the animal groups Hymenoptera, Monogononta, Homoptera, Thysanoptera, Coleoptera and Acarida. If one locus at least is allozygous in the mother, the male offspring will not receive a copy of all maternal genes.)

The two categories of asexual reproduction described by Williams (1975) are: primitive and derived. The primitive category includes mitotic fission and vegetative reproduction. The derived category includes polyembryony of higher animals, and possibly some growth-related spread of higher plants and fission in some worm groups. It also includes amictogametic forms of reproduction. These encompass '. . . any development from ova, spores etc. that preserve the maternal genotype . . .'

Williams appears to require that either nuclear fusion or, in its absence, the production of genetically variable progeny, must occur before a reproductive process can be considered to be sexual. This definition is therefore part mechanistic (in its emphasis on nuclear fusion) and part functional (in its emphasis on genetic variation). His definition of asexual reproduction includes any process that produces progeny whose genotypes are copies of the maternal genotype, even if the propagules that initiate the progeny are multicellular.

Bell (1982: 43–44) disagrees with much of Williams' classification system, criticizing it for, among other things, reflecting

... not so much the nature of the processes involved as the author's opinions about the way in which they have evolved; his classification thus hinges on a particular interpretation of sexual processes, and cannot very well be used to discuss other interpretations ...

Bell's own system of classification includes three main categories of reproduction: amictic, partially mictic and mictic (Bell, 1982: 36–37). Here 'mixis', which is synonymous with 'sex', describes a process

... which changes the relationship between different elements (genes or linkage groups) of the genome, with or without the introduction of completely novel genetic material ... (Bell, 1982: 20)

Under this scheme, amictic (asexual) reproductive processes are mitotic. Syngamy (nuclear fusion or fertilization) is absent. They include colonial and clonal proliferation, of which the latter encompasses vegetative propagation, polyembryony and apomixis. Partially mictic processes are those in which syngamy or meiosis are restricted to one sex. It encompasses arrhenotoky. Mictic (sexual) processes are those which include both a reductional meiosis and the restoration of diploidy by syngamy. Two categories are recognized: automictic and amphimictic. Automictic processes involve fusion between nuclei derived from the same zygote. Amphimictic processes involve the syngamy of gametes of different gender derived from different zygotes. Among multicellular eukaryotes, at least, amphimixis is synonymous with allogamy (outcrossing).

Bell's scheme differs from that of Williams' in that it classifies reproductive processes in terms of their mechanisms rather than in terms of a combination of mechanisms and function. According to Bell, sexual reproduction involves nuclear fusion, asexual reproduction does not. Because it is a mechanistic definition, a process will always qualify as either sexual or asexual. Thus from two recent and influential texts we have two very different opinions of which reproductive processes are sexual and which asexual. Bell views sexual reproduction as a process – specifically one that involves meiotic reduction and the fusion of nuclei. Williams is more inclined to see it as an effect – the production of genetically variable progeny, by almost any process. Most of the definitions to be found in the literature veer towards one or other of these views, although some uncomfortably attempt to encompass both.

A major problem with functional definitions is that we do not yet understand the primary function of sex. It is for this very reason that this whole area has so attracted the attention of biologists! We know many of the consequences of different forms of reproduction but we are still very unsure about which ones are the primary reasons for their maintenance and which are merely secondary, albeit possibly very advantageous, consequences. For example, we know that allogamy can generate progeny with variable and novel genotypes but, through its association with meiosis, it can also effect DNA repair. It can also lead to reduced sib competition and to resistence against pathogens, and it can retard the progress of Muller's Ratchet. These issues are described in detail in the review edited by Michod and Levin (1988), which provides a very readable account of the present state of play. The uncertainty about which, if any, of these consequences of allogamy is primarily responsible for its maintenance is still very real, after two decades of intense and invigorating research and debate. Because of the lack of a stationary and easily visible target, functional definitions of sex tend to lack any clear direction. Until such a target appears, a mechanistic definition is to be preferred.

The definition I favour is very stark indeed. It is simply that sexual reproduction involves the initiation of an individual from a single cell, which must be a zygote or zygote equivalent, whose nucleus is formed by syngamy. This definition differs from Bell's in that it does not require the fusing nuclei to be the products of a reductional meiosis or for syngamy to restore the parental chromosome number. Of course, both of these will typically occur, but we know of many instances of fusion between meiotically unreduced gametes which, in plants at least, have often given rise to successful polyploid lineages (Harlan and de Wet, 1975; Lewis, 1980). I would not wish to exclude such fusions as sexual. I will define asexual reproduction as the initiation of an individual from a single cell, which must be a zygote equivalent, whose nucleus is not formed by syngamy.

Given the differences of opinion that exist about the definitions of sexual and asexual reproduction, it follows that some processes (e.g. parthenogenesis) have no consistent status as sexual or asexual phenomena. This problem will be discussed in the next chapter.

Chapter 2

Patterns of reproduction in bryophytes and tracheophytes

2.1 INTRODUCTION

One of the greatest pleasures to be obtained from the study of organisms is the kindling of an awareness of the seemingly limitless variety that has been generated from a handful of basic life history blueprints. Multicellular eukaryotic plants are a case in point. The basic life history is mundane in its constancy – a zygote divides mitotically to form a sporophyte, which differentiates meiospores by meiosis, each of which divides mitotically to form a gametophyte, which differentiates gametes, which fuse in pairs to form zygotes. But within these confines great ontogenetic adventures have been had. One aspect of life history that has been notably fluid in its response to the forces of selection has been that of reproduction. The evolution of the paraphernalia of reproduction and of reproductive strategies is an epic play, although our knowledge of many of its scenes is incomplete, with crucial passages still to be dredged and sifted from the fossil record. Happily, most of the information needed to set the background for this discussion is readily available. It is the purpose of this chapter to present it, albeit briefly.

An unavoidable though restrictive consequence of progress in biology has been the development of a specialized descriptive vocabulary. To readers not very familiar with plant reproductive biology, I am afraid that this chapter will provide yet another supplement to this. I ask you to grit your teeth and to submit, if not gracefully then at least robustly, to the requirement to become familiar with it. This chapter, although it says little that has not already been said elsewhere, is an essential part of this book, setting the background for much that follows. To readers who are familiar with plant reproductive biology, may I suggest that this chapter is still worth considering in detail, as I have felt it necessary both to bring together information that is not normally juxtaposed, and to stress areas that are not usually stressed.

It is necessary to approach this chapter's goal from various directions, for although great differences in reproductive biology separate major taxa, there are also aspects that unite them, having changed slowly, if at all, through phylogeny. The production of eggs within archegonia falls into this category. As I will be arguing in Chapter 3 that the presence and the type of archegonia have had a considerable determining effect on the capacities of plants to adopt particular forms of asexual reproduction, this organ will be described. But before I do this I will first review the different patterns of reproduction – both sexual and asexual – that are exhibited by oogamous eukaryotes. In doing this I will be providing the background against which discussions of the capacity of different major taxonomic groups to evolve particular patterns of reproduction can be developed. Following this review and the description of archegonia, a brief overview will be given of some of the major changes in reproductive biology that have characterized plant phylogeny. Attention will be focused on characteristics that have a direct bearing on the frequency and distribution of different types of asexual reproduction, such as homospory, heterospory and the seed habit. This will be followed by an account of relevant aspects of reproductive biology in the major extant classes of the Bryophyta and Tracheophyta.

2.2 PATTERNS OF REPRODUCTION IN OOGAMOUS EUKARYOTES

The central role in reproduction of a zygote or zygote equivalent provides a means by which different types of reproductive processes can be recognized. A classification system can be developed based on the origin of the nucleus of this cell. This is attempted in Fig. 2.1, which identifies 12 major methods of reproduction in oogamous eukaryotes. The key presented in Fig. 2.1 would need only minor modifications to make it applicable to all eukaryotes. Some of the forms of reproductive processes described are restricted to plants. These are noted in appropriate parts of the key. The reasons for their restriction to plants need not concern us here, as they have no bearing on the present discussion. However, they will be discussed later in this chapter and in Chapter 3.

1 Nucleus of the cell that initiates a new individual during reproduction formed by the fusion of nuclei (2)

 Nucleus not formed by fusion (5)

2 Fusing nuclei produced by the same parent (3)

 Fusing nuclei produced by different parents (allogamy)

3 Fusing nuclei produced by the same meiosis (4)

 Fusing nuclei produced by different meioses (autogamy)

4 Fusing nuclei are from cells differentiated as gametes (non-parthenogenetic automixis)

 Fusing nuclei are meiotic tetrad nuclei, the nucleus formed by fusion subsequently becoming the nucleus of an egg cell which develops parthenogenetically (parthenogenetic automixis)

5 Nucleus a product of a reductional meiosis (6)

 Nucleus not formed by a reductional meiosis (9)

6 Meiosis preceded by an endomitosis, meiosis restores the parental number of chromosomes (7)

 Meiosis not preceded by an endomitosis, the nucleus contains only half the parental number of chromosomes, and is the nucleus of an egg cell that develops parthenogenetically (8)

7 Nucleus is that of an egg cell which develops parthenogenetically (*Allium* scheme of parthenogenesis)

 Nucleus is that of a somatic cell of a plant gametophyte stage (Döpp–Manton scheme of agamospory)

8 Parthenogenesis not followed by fusion of cleavage division nuclei, so that the organism is haploid (non-replicative haploid parthenogenesis)

 Parthenogenesis followed by fusion of cleavage division nuclei, so that the organism is diploid (replicative haploid parthenogenesis)

9 Nucleus is that of an egg cell which develops parthenogenetically (10)

 Nucleus is that of a plant somatic cell (11)

10 Egg cell derived from generative cell (generative apomixis)

 Egg cell derived from somatic cell in ovule of flowering plant (aposporous apomixis)

11 Somatic cell is part of vegetative body of gameophyte stage (Braithwaite scheme of agamospory)

 Somatic cell is part of ovule of flowering plant (adventitious embryony)

Figure 2.1 A dichotomous key identifying the major methods of reproduction in oogamous eukaryotes. See the main text for a full discussion of the key and of the terminology used.

A major dichotomy separating reproductive processes concerns whether or not the nucleus of the cell that initiates a new individual is formed by the fusion of nuclei. If it is, the origin of the fusing nuclei determines the type of reproductive process. If the fusing nuclei are produced by different individuals, the process is one of outcrossing (allogamy). If they are produced by the same individual, the process is one of selfing. Two categories of selfing can be recognized: autogamy and automixis. These terms are applied inconsistently, as can be seen from Table 2.1. Selfing will be defined here as autogamy if the nuclei are derived from different meioses, and as automixis if they are derived from the same meiosis. The fusions in autogamous selfing are typically between the nuclei of fully differentiated male and female gametes. But two types of automixis can be recognized, only one of which involves fusion between such gametes. This type is found in bryophytes and pteridophytes where, with few exceptions, each member of a meiotic tetrad differentiates as a meiospore, with either each meiospore giving rise to a cosexual gametophyte (ferns and some bryophytes) or with two of the meiospores giving rise to male gametophytes and two to female gametophytes (some bryophytes). Consequently, each meiosis generates both male and female gametes, which can fuse in self-compatible individuals. The other type of automixis (which does not involve fusion between fully differentiated gametes) has been observed in a number of animal taxa (Bell, 1982). Here, two of the tetrad nuclei produced during a meiotic division in a female fuse, and the fused product becomes the nucleus of an egg that develops parthenogenetically. I will describe this type of automixis as parthenogenetic automixis and the other type (which involves fusion between fully differentiated gametes) as non-parthenogenetic automixis.

The product of fusion in allogamy, autogamy and non-parthenogenetic automixis is a zygote. That in parthenogenetic automixis is a zygote equivalent. Consequently, all four processes attract the status of sexual reproduction.

During a number of reproductive processes, the nucleus of the cell that initiates the new individual is not formed by the fusion of nuclei. Consequently, these processes qualify as asexual reproduction, and the initiating cells attract the status of zygote equivalents. The initiating cell may be an egg cell which develops parthenogenetically, or a somatic cell which develops mitotically to form an embryo. The former process, which will be described first, is found in both animals and plants, the latter only in plants.

Five categories of asexual reproduction involving parthenogenesis can be distinguished on the basis of differences in the pattern of egg production or division. Three are associated with a reductional meiosis. In the first of these, meiosis is preceded by an endomitosis.

Table 2.1 Synonyms of, and comments about, the processes of reproduction identified in Fig. 2.1. The list of synonyms is not meant to be exhaustive, and the reference list gives only the first example, or recent examples, of their use

Terms used in this book	Synonym	Comment	References where used
Adventitious embryony	Nucellar embryony		
Allium scheme of parthenogenesis	Apomixis		Hakansson and Levan (1957) Nogler (1984)
Allogamy	Amphimixis	Often simply described as 'outcrossing'	
Apogamy	Apogamety		Gustafsson (1946–47) Richards (1986)
Aposporous apomixis	Apoarchespary		Khokhlov (1976)
Autogamy	Automixis	Often simply described as 'self-fertilization' but, in the references cited, is described as a category of automictic reproduction, 'automixis' encompassing any reproductive process in which nuclei derived from the same zygote fuse	Drebes (1977) Bell (1982)
Braithwaite scheme of agamospory	*Asplenium aethiopicum* system of apomixis	'Agamospory' and 'apomixis' are used interchangeably in the literature	Walker (1966)
	Mehara–Singh scheme of apomixis		Klekowski (1973)
Döpp–Manton scheme of agamospory	Normal type of apomixis	'Agamospory' and 'apomixis' are used interchangeably in the literature	Walker (1966) Mogie (1986a)
Generative apomixis	Aposporous apomixis	Usually described as 'diplosporous apomixis' in plants and as 'apomixis' in animals	Khokhlov (1976)

Non-parthenogenetic automixis	Many texts make no terminological distinction between this and autogamy, describing both simply as instances of self-fertilization; the references given provide a few examples	Fritsch (1965) Crew (1965) Ingold (1973) Klekowski (1973) Round (1973) Sporne (1975) Bold and Wynne (1978) Fincham et al. (1979) Page (1979)
Non-replicative haploid parthenogenesis	Haploid parthenogenesis, reduced, generative parthenogenesis	Nogler (1984)
Parthenogenetic automixis	Matromorphy	Eenink (1974) Asker (1980)
	Parthenogamy	Crew (1965)
Replicative haploid parthenogenesis	Automixis	Pijnacker (1969) Maynard Smith (1978)

Consequently, the reductional meiosis generates an egg that exhibits the same number of chromosomes as the mother. I will describe this process, which occurs in a number of animal taxa, including vertebrates, and in a few plants, including *Allium nutans*, as the *Allium* scheme of parthenogenesis. In the second and third of these three categories the reductional meiosis is not preceded by an endomitosis. Consequently, the egg exhibits only half the number of chromosomes as the mother, and reproduction is by haploid parthenogenesis. However, the two categories differ in the events that immediately follow the mitotic division of the egg (the first cleavage division). This division may be normal so that the new individual is haploid (this is how males are produced in arrhenotokous taxa) or it may be endomitotic (i.e. the cleavage division nuclei fuse) so that the new individual achieves diploidy immediately after its initiation (it is haploid at initiation as the egg, being a zygote equivalent, is the first cell of the new individual in parthenogenetically reproducing taxa). This process has been observed as an occasional or regular phenomenon in some coccid bugs, members of the Arachnida (scorpions, mites, spiders etc.) the stick insect *Bacillus rossius*, the goldfish *Carassius auratus gibelio*, and members of the cruciferous flowering plant genus *Brassica* (Pijnacker, 1969; Maynard Smith, 1978; Bell, 1982; Eenink, 1974). I will describe this process as replicative haploid parthenogenesis, and the former process (in which the fusion of cleavage division nuclei does not occur) as non-replicative haploid parthenogenesis; note that in these descriptions the object experiencing or avoiding replication is the genome of the new individual.

The two remaining categories of asexual reproduction involving parthenogenesis do not involve a reductional meiosis. In the first of these the egg is derived from a generative cell (the oocyte in animals, the megaspore mother cell, also described as the embryo sac mother cell in flowering plants) by mitosis or by a restitutional meiosis. Consequently, it exhibits the parental (sporophytic) number of chromosomes. This process, which is exhibited by many animal and plant taxa, is known simply as apomixis when it applies to animals but as diplosporous apomixis (abbreviated to 'diplospory') when it applies to plants. It has recently been suggested that a single term – generative apomixis – should be used to describe it in both groups (Mogie, 1988). The remaining category of parthenogenesis is exhibited by some flowering plants. The egg is derived mitotically from a somatic cell in the ovule, instead of from the generative cell. It consequently exhibits the parental number of chromosomes. This reproductive process is aposporous apomixis (abbreviated to 'apospory').

As mentioned earlier, a number of plants reproduce asexually from cells other than egg cells. Some flowering plants differentiate embryos directly from somatic cells of the ovule (which is part of the sporophyte stage) in a process called adventitious (or nucellar) embryony. A similar process is found in the ferns, although here the somatic cell that gives rise to the embryo is a vegetative cell of the gametophyte stage. The spore that initiates the gametophyte may result from a restitutional meiosis, in which case reproduction is by the Braithwaite scheme of agamospory (Braithwaite, 1964), or from a reductional meiosis of a cell that has undergone a premeiotic endomitosis, in which case reproduction is by the Döpp-Manton scheme of agamospory (Klekowski, 1973). In both cases, the spore exhibits the sporophytic number of chromosomes.

Some of the terminology used here to describe the twelve reproductive processes found in oogamous eukaryotes is used inconsistently in the literature. In the zoological and general literature, processes involving a premeiotic endomitosis, or the fusion of cleavage division nuclei, are often described as automictic. This is not the case within the botanical literature, where they are often described as apomictic. I have removed them from both of these categories, classifying them instead, respectively, as the '*Allium* scheme of parthenogenesis' and 'replicative haploid parthenogenesis'. I have adopted this system of classification because I believe that it is useful to distinguish these two processes and the Döpp–Manton scheme of agamospory from apomictic processes and that it is important to distinguish them from automictic processes. (I speak fervently as a convert, having classified them as automictic in a previous publication (Mogie, 1986a).) Apomixis is not characterized by an episode of nuclear fusion (at least if the reforming of the nucleus at the end of a restitutional meiotic division is not interpreted as a fusion event), whereas these three processes are. However, fusion in these does not occur at the point during reproduction when a new individual is initiated. I prefer to reserve the term 'automixis' for those reproductive processes in which fusion does occur at this point. Thus the nucleus formed by endomitotic fusion in the *Allium* and Döpp–Manton schemes contains twice as many chromosomes as the nucleus of the cell that initiates the new individual – the endomitotically produced nucleus is part of a maternal cell, not the initiating cell. The initial cell of the new individual is a meiotic derivative of this cell. Similarly, the nucleus produced endomitotically during replicative haploid parthenogenesis is not the nucleus of the initiating cell of the new individual, but is a mitotic derivative of this cell. The nucleus of the initiating cell is the prefusion (pre-endomitosis), postmeiotic, haploid nucleus of the parthenogenetically developing egg. That is, reproduction is achieved by haploid parthenogenesis, diploidy by growth.

Another aspect of the classification system proposed that may be controversial is the terminology itself. Most reproductive processes are described by several synonyms. The most commonly used of these are listed in Table 2.1. All I can say in defence of the ones chosen for use here are that they are the ones I am most comfortable with. However, my use of 'autogamy' and 'automixis' may irritate word-smiths, as I do not use them in their literal sense. With respect to their literal meanings, 'automixis' should describe the fusion of nuclei derived from the same zygote, with 'autogamy' being identified as a special case of automixis – that of the fusing nuclei being enclosed within gametes. However, if used in this way, no clear distinction is made between fusion of nuclei of the same meiosis or of different meioses – intermeiotic selfing is always autogamic, but intrameiotic selfing is predominantly non-autogamic in animals but predominantly autogamic in plants. As the genetic consequences of intra- and inter-meiotic fusion can be very different (these are summarized in Table 2.2), it is important to distinguish between them. New terms could be introduced, but these would add to the already burdensome termin-ology associated with reproduction. Rather than do this, I have opted to use 'automixis' to describe intrameiotic fusion, and 'autogamy' to describe intermeiotic fusion.

Before concluding this section it is worth noting that only two of the eight asexual processes described are guaranteed (barring mutation) to pass on the maternal genotype intact to progeny. These two pro-cesses are aposporous apomixis and adventitious embryony. Whether the other processes do this is dependent on several factors. Thus if the mother is allozygous at some loci, both replicative and non-replicative haploid parthenogenesis will produce genetically variable progeny. Likewise, whereas both generative apomixis and the Braithwaite scheme of agamospermy will preserve the maternal genotype if mei-osis is replaced by mitosis, or if the restitutional meiosis is completely asyndetic, they will produce genetically variable progeny if there is even a small amount of recombination during the restitutional mei-osis. Similarly, both the *Allium* scheme of parthenogenesis and the Döpp-Manton scheme of agamospermy include a recombinational mei-osis, but this will only result in genetically variable progeny if non-sister, rather than sister, chromosomes pair. (In a tetraploid cell formed by endomitosis, non-sister chromosomes are homologous. They would normally pair during meiosis if the cell was diploid. Sister chromoso-mes are replicas of each other, being derived from different chromatids of the same mitotically dividing chromosome. In a normal mitosis they pass in to sister cells. In an endomitosis they remain in the same cell. Recombination between sister chromosomes will not result in genetic change.) Thus asexual reproduction is not necessarily a mech-anism for preserving maternal genotypes.

Table 2.2 The status of the 12 reproductive processes described in the text and in Fig. 2.1, according to Williams (1975), Bell (1982) and this book, and a summary of their genetic consequences. The references describe the latter in more detail. S = sexual; A = asexual; V = of variable status. A* = Bell (1982) discusses haploid parthenogenesis in arrhenotokous taxa. In these, males are produced asexually by haploid parthenogenesis but females are produced sexually. Bell consequently describes arrhenotoky as partially sexual

Reproductive process	Status according to Williams/Bell/Mogie	Genetic consequences
Allogamy	S/S/S	Progeny genetically variable, rate of inbreeding reduced, high levels of heterozygosity maintained, Muller's Ratchet unimportant (Maynard Smith, 1978; Falconer, 1981).
Autogamy	S/S/S	50% reduction in the proportion of heterozygous loci per individual per generation leading rapidly to highly homozygous genotypes and genetically invariable progeny (Falconer, 1981).
Parthenogenetic automixis	S/S/S	Homozygosity increases, with the pattern of increase being influenced by the pattern of fusion. Thus, during the first meiotic division, the generative cell gives rise to two nuclei (A and B), each of which gives rise to two derivatives (a, a and b, b) during the second meiotic division. If sister nuclei fuse (i.e. a × a or b × b) homozygosity may increase between the centromere and the first chiasma but heterozygosity will be maintained for distal loci. If non-sister nuclei fuse (i.e. a × b) heterozygosity will be preserved for loci situated between the centromere and the first chiasma but may be lost for distal loci (White, 1973; Maynard Smith, 1978; Bell, 1982).
Non-parthenogenetic automixis	S/S/S	The same as for parthenogenetic automixis. However, additionally, in plants with cosexual gametophytes, the gametes that fuse can be mitotic derivatives of the same meiotic tetrad nucleus. This type of fusion will lead to 100% homozygosity in a single generation (Klekowski, 1973).
Replicative haploid parthenogenesis	V/S/A	If the mother is heterozygous, each initiating cell will be genetically distinct, but each offspring will be 100% homozygous following endomitotic fusion of cleavage division nuclei.

Non-replicative haploid parthenogenesis	V/A*/A	If the mother is heterozygous, each initiating cell will be genetically distinct. This is very likely in arrhenotokous taxa, as the diploid females tend to be produced allogamously (Bell, 1982).
Allium scheme of parthenogenesis, and Döpp–Manton scheme of agamospory	V/S/A	Heterozygosity will be maintained, but the progeny will be genetically identical to each other and to the mother, if only sister chromosomes pair. However, if non-sister chromosomes pair (in bivalents or with sister chromosomes in tetravalents) segregation can occur and levels of heterozygosity will be reduced (Klekowski, 1973; Bell, 1982).
Braithwaite scheme of agamospermy, and Generative apomixis	V/A/A	If there is recombination during the restitutional meiosis, segregation can occur and levels of homozygosity will increase. If the restitutional meiosis is completely asyndetic, or if it is replaced by a mitotic division, levels of heterozygosity will be maintained, but the progeny will be genetically identical to one another and to the mother.
Aposporous apomixis, and Adventitious embryony	A/A/A	These are mitotic processes. Consequently, the progeny are identical to each other and to the mother.

It is worth considering the asexual processes further to see which are most likely to generate genetically variable progeny. Arrhenotokous taxa appear to be mostly allogamous (Bell, 1982). As a consequence, diploid mothers will be heterozygous and the sons they produce by non-replicative haploid parthenogenesis will be genetically variable. However, an organism produced by replicative haploid parthenogenesis will be homozygous at all loci. If it in turn reproduces by replicative haploid parthenogenesis its progeny will be genetically identical. Animal taxa reproducing by the *Allium* scheme of parthenogenesis are generally considered to undergo meiosis in which only sister chromosomes pair (Maynard Smith, 1978), resulting in the maternal genome being inherited intact. The Döpp–Manton scheme of agamospermy is analogous in many respects to the *Allium* scheme of parthenogenesis. However, controversy surrounds the issue of whether sister or non-sister chromosomes pair during meiosis (Klekowski, 1973; Lovis, 1977). Finally, it is probable that the restitutional meiosis in most organisms reproducing by the Braithwaite scheme of agamospory, or by generative apomixis, is wholly asyndetic. However, there are a number of reports of chromosome pairing in generatively apomictic plants (Gustafsson, 1946–47). Generative apomixis cannot therefore be viewed as a process that is guaranteed to preserve the maternal genome.

Of course, the way the 12 reproductive processes described in Fig. 2.1 have been partitioned into four sexual and eight asexual types reflects the way I distinguish between sexual and asexual reproduction. It should be clear from Chapter 1 that other biologists would distribute them differently. This is illustrated in Table 2.1 which, as well as providing a brief description of the genetic consequences of the different reproductive processes, divides them into sexual and asexual processes according to my interpretation of the views of Williams (1975) and Bell (1982), which are summarized in Chapter 1.

2.3 SEX ORGANS AND GAMETOPHYTE ONTOGENY IN BRYOPHYTES AND TRACHEOPHYTES

Many of the phylogenetic relationships within and between bryophytes and tracheophytes are still unclear (Duckett and Renzaglia, 1988). However, there is little doubt that these groups are derived from a common ancestral division – the Chlorophyta (green algae). But there is some argument about whether they each arose independently from this division, or whether the Tracheophyta gave rise to the Bryophyta. A basic phylogenetic sequence is depicted in Fig. 2.2, with the main phylogenetic moves being described in the legend to this figure. The present uncertainty about the relationships between different taxa is evident from this; it will not be resolved until more information is prised from the fossil record. Many of the taxa depicted in Fig. 2.2 are known only from their fossil remains, and little is known of their reproductive biology. Attention will therefore be focused on some of their living relatives. These comprise members of the bryophyte classes Hepaticopsida (liverworts), and Bryopsida (true mosses), and of the tracheophyte classes Filicopsida (ferns), Gymnospermopsida, Gnetopsida and Angiospermopsida.

Characteristics common to these are that reproduction is oogamous, the zygote is retained by the gametophyte until at least a sizeable embryo is formed, and meiospores (which are produced by the sporophyte) and gametes (which are produced by the gametophyte) are produced in multicellular, protective organs. But around these common themes there is much variation, some of which has had a determining effect on the incidence and types of asexual reproduction in different classes. This variation will be explained here. However, the main purpose of this chapter is to provide a description of the reproductive processes that are found among plants. Thus the interpretation of the variation observed will be postponed until later chapters. The description will commence in this section with a commentary on the processes by which gametes and embryos are produced.

In all bryophytes, and in all tracheophytes except flowering plants and the gymnospermous genera *Gnetum* and *Welwitschia*, the organ within which the egg and embryo develop is the archegonium. Each archegonium usually contains a single egg, but a gametophyte may produce many archegonia. Differences are found within and between classes in key aspects of archegonium development. It is worth considering these as I will be arguing in Chapter 3 that they have had a profound effect on the capacity of some classes to evolve asexual reproduction and of others to evolve particular forms of asexual reproduction. The archegonium is a well-differentiated and more-or-less flask-shaped structure which is embedded in the vegetative tissues of the gametophyte. It comprises a swollen base (the venter), and a more elongated slender neck which may protrude above the surface of the gametophyte. It consists of an axial row of cells surrounded by a sterile jacket of cells. Most commonly, the axial row forms a single series, which is derived from mitotic divisions of a 'central cell'. In bryophytes, and in ferns and fern allies (which will be described collectively as pteridophytes), the central cell divides mitotically to form a primary canal cell and a primary ventral cell. Between one and 16 neck canal cells are produced mitotically from the primary canal cell. The primary ventral cell divides mitotically to produce a ventral canal cell and an egg (Foster and Gifford, 1959). The basic structure of the mature archegonium, and its development, are illustrated in Fig. 2.3(a–c). However, numerous deviations from this pattern are found in the gymnosperms. These involve reductions in the number of mitotic divisions undertaken during the production of the axial row of cells.

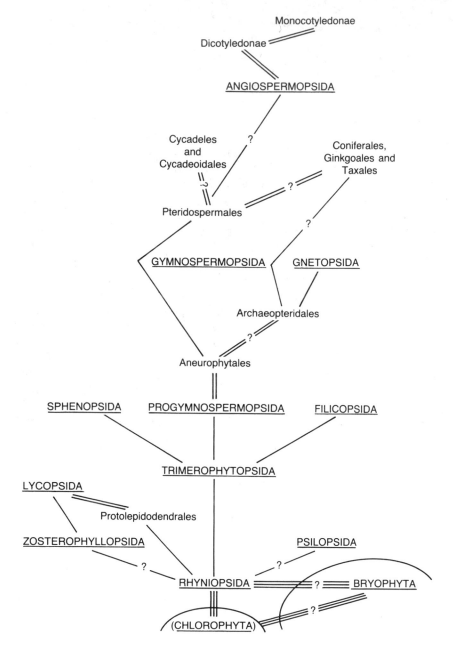

Thus, neck canal cells are absent. In some taxa, rather than dividing to form a primary canal cell and a primary ventral cell, the central cell divides instead to produce a binucleate structure comprising a ventral canal nucleus and an egg (e.g. *Cephalotaxus*, *Ephedra*, Taxodiaceae and Cupressaceae, Fig. 2.3a,d). In others, most notably the Pinaceae, the central cell divides to from a ventral canal cell and an egg cell (Fig. 2.3a,e). In *Taxus*, *Torreya taxifolia* and *Widdringtonia cupressoides*, the central cell does not divide, but functions directly as the egg (Fig. 2.3a,f); these taxa therefore lack both neck canal cells and a ventral canal cell.

Clearly, the egg in gymnosperms is not the homologue of the pteridophyte egg but is a mitotic precursor of it. Nor is homology maintained within the gymnosperms, the egg of some taxa being homologous to

---

Figure 2.2 The phylogenetic relationships between the divisions Bryophyta and Tracheophyta, and between the 11 classes of the Tracheophyta. Classes arc depicted in upper case and are underlined. The Monocotyledonae and Dicotyledonae are subclasses. All other taxa depicted are orders. Phylogenetic sequences run from bottom to top. Single lines link classes in a sequence. Double lines link an order or subclass to its class. Treble lines link divisions. For example, the ancestral tracheophyte class Rhyniopsida is derived directly from the division Chlorophyta. Within the Tracheophyta, thc ordcr Protolcpidodendrales is a member of the class Lycopsida, the order Cycadeles of the class Gymnospermopsida. Taxa below the level of classes are used to clarify phylogenetic sequences. For example, the Lycopsida appear to be derived in part from the Zosterophyllopsida and in part from the Rhyniopsida. A (?) placcd on a line denotes a tentative phylogenetic link. This figure is based on discussions in Stewart (1983), Beck and Wight (1988), Doyle and Donoghue (1987) and Crane (1985, 1988).

The divisions Bryophyta and Tracheophyta are both descended from the division Chlorophyta (green algae). However, there is still considerable debate about whether they arc cach descended independently from this taxon, or whether thc Bryophyta is descended from a member of the Rhyniopsida – the most ancient tracheophyte class. The Rhyniopsida gave rise directly to the Zosterophyllopsida, Psilosida, Lycopsida and Trimerophytopsida. The Trimerophytopsida holds a pivotal position in tracheophyte evolution, for from it evolved all the tracheophyte taxa that are currently most abundant, including the ferns (Filicopsida) and fern allies, the gymnosperms and the flowering plants – the latter two groups comprising the seed plants. The seed and non-seed lines probably emerged independently from the Trimerophytopsida. The seed plants can be traced from the Progymnospermopsida. The exact relationship between the non-flowering seed plants and the flowering seed plants is in some doubt. Both may have evolved as independent lines from the Progymnospermopsida, with the angiosperms evolving from a cycadeles or cycadeoidales type ancestor, as depicted in the figure, or they may be more directly related. The immediate ancestor of the angiosperms has not been positively identified. The original angiosperms were almost certainly dicotyledonous, with an early dichotomy resulting in the divergence of the Dicotyledonae and the Monocotyledonae.

mitotic precursors of that of others. (This comparison is valid. It can be seen from Fig. 2.2 that the pteridophytes, gymnosperms and flowering plants share tracheophyte ancestors in common.) Thus there has been a tendency during tracheophyte phylogeny for egg initiation to be brought forward to earlier stages of archegonium development. This appears to be a clear example of progenesis. It is most notable during the part of phylogeny associated with the emergence of the seed habit. It is therefore of some interest that the gymnospermous

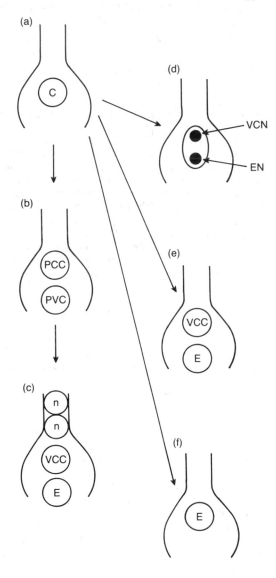

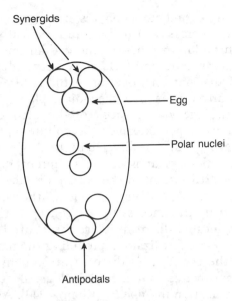

Figure 2.4 The distribution of the eight nuclei of the 'Normal' or 'Polygonum-type' embryo sac (female- or mega-gametophyte) of flowering plants. This is the most common type found in this class. It is formed by the successive mitotic divisions of the megaspore, and comprises two synergids, an egg, two polar nuclei and three antipodals, all of which are haploid. The extent to which the eight nuclei are enclosed within their own cell walls varies between taxa, although it is common for a cell wall to form around the egg nucleus.

genera *Welwitschia* and *Gnetum* do not produce organs that resemble archegonia. Nor do flowering plants, the entire female gametophyte comprising only a few (most frequently eight) nuclei, one of which functions as the egg (Fig. 2.4). It is a matter for debate, the pursuance of which will be left to the next chapter, whether archegonium production has actually been lost by flowering plants, or merely disguised by a further progenetic shift.

Figure 2.3 The development of the axial row of an archegonium in the bryophytes and tracheophytes. The pathway in bryophytes and ferns involves steps (a), (b) and (c). All divisions are mitotic. The central cell (C) in (a) divides to produce the primary canal cell (PCC) and the primary ventral cell (PVC). The primary canal cell initiates one or more divisions to give two or more neck canal cells (n) and the primary ventral cell divides to produce the ventral canal cell (VCC) and the egg (E). The pathway in gymnosperms involves steps (a) and (d), (a) and (e) or (a) and (f). Thus in the first pathway, the nucleus of the central cell divides to give the ventral canal nucleus (VCN) and the egg nucleus (EN) in (d). In the second pathway, the central cell divides to give the ventral canal cell (VCC) and the egg cell (E) in (e). In the third pathway the central cell behaves as the egg cell.

In bryophytes and pteridophytes, male gametes are produced within antheridia. Whereas each archegonium typically produces only a single egg, an antheridium produces more than one male gamete and can produce many thousands. Obvious antheridia are not produced by gymnosperms and angiosperms, which differentiate male gametes from specialized cells. In archegoniate plants, the male gamete gains entry to the egg via the neck of the archegonium. Male gametes may be motile spermatozoids (bryophytes, pteridophytes and some gymnosperms) or may be directed to the egg by a pollen tube (flowering plants and most gymnosperms). Different taxa vary in the events leading to fertilization. In those that produce archegonia with neck canal cells, these cells break down, thereby providing the male gametes with direct access to the venter of the archegonium where the egg is located. In many taxa, the ventral canal cell also breaks down just before fertilization, providing the male gametes with direct access to the egg. This is the case in bryophytes, pteridophytes and most gymnosperms, but the ventral canal cell nucleus can persist during fertilization in *Ephedra* (Khan, 1943). Although more than one male gamete may enter an archegonium, only one fertilizes the egg.

The embryo in archegoniate plants starts to develop immediately after fertilization. There is no resting period. This can be contrasted with the situation in flowering plants, and in the non-archegoniate gymnospermous genus *Gnetum*, where the zygote may remain dormant for a considerable period. The zygote of *Gnetum* has been observed to rest for 15 days (Vasil, 1959). Those of flowering plants also typically rest for several days, some for much longer. For example, that of *Colchicum autumnale* rests for up to 5 months (Favre-Duchartre, 1984).

The embryo develops *in situ* in all bryophytes and tracheophytes, obtaining protection and nutritional support from the archegonium (when present) and, in non-flowering plants, from the surrounding gametophytic tissue. Nutritional support in flowering plants is primarily provided by the endosperm.

The endosperm results from the fusion of one or more (usually two) maternal gametophyte nuclei (polar nuclei, Fig. 2.4, but see Battaglia (1987) for an alternative terminology) with a male gamete that is genetically identical to the gamete that fertilized the egg; that is, the two male gametes that between them fertilize the egg and the polar nuclei are mitotic derivatives of a pregametic cell of the paternal gametophyte, both being transported to the maternal gametophyte by the pollen tube. The endosperm initial resulting from this fusion divides mitotically to give rise to the endosperm, which acquires resources from the maternal sporophyte, which are passed, along with growth regulating substances, to the developing embryo. This phenomenon of double fertilization is unique to flowering plants.

With the exception of the fern *Anogramma leptophylla* (Mehr and Sanhu, 1976), the embryo (sporophyte) of all extant tracheophytes eventually develops to the stage where it no longer requires the protection and support of the maternal tissues. It achieves independence, and the sporophyte is the dominant stage of the life cycle. The bryophyte sporophyte is permanently dependent on the maternal gametophyte, which is the dominant stage of the life cycle.

The meiospore which gives rise to the gametophyte is the product of the meiotic division of a generative cell of the sporophyte. Generative cells, and hence meiospores, are produced within specialized multicellular structures identified as capsules in bryophytes and as sporangia in tracheophytes. At maturity, the meiospore-derived gametophyte may produce both male and female gametes, or only one type of gamete. The former behaviour is found in approximately 50% of bryophyte species and in most pteridophyte species. The latter behaviour is found in flowering plants, gymnosperms and the remaining bryophyte and pteridophyte species.

This distinction between a gametophyte that can typically produce both male and female gametes and one that can produce either male or female gametes but never both marks a major dichotomy within the Tracheophyta. Homosporous taxa produce the former type of gametophyte, heterosporous taxa the latter type. It is clear that heterospory evolved from homospory (Turnau and Karczewska, 1987; Haig and Westoby, 1988, 1989). Each meiosis in homosporous taxa results in four viable meiospores which are dispersed following the dehiscence of the sporangium. The meiospores are essentially equal in size, although size differences have been reported which do not, however, follow any consistent pattern. Each meiospore gives rise to a free-living gametophyte that supports itself either through photosynthesis, or through an association with a mycorrhizal fungus, and which generally comprises many cells and produces numerous antheridia and archegonia. However, antheridia are typically initiated before archegonia, and gametophytes which are produced by very small meiospores, or which fail to develop properly, often succeed in producing functional antheridia but fail to produce archegonia. This is a phenotypic response rather than a genetically determined condition.

Male meiosis in heterosporous taxa is akin to meiosis in homosporous taxa. The four meiospores produced by each meiosis are described as microspores. These divide mitotically to form male gametophytes (microgametophytes or pollen) which are released following the dehiscence of the sporangium (described as the microsporangium). In contrast, each female meiosis in heterosporous taxa generates only a single functional meiospore (some members of the pteridophyte genus *Selaginella* are exceptions, Sporne, 1975). In most taxa, this contains only one of the four tetrad nuclei, the other three degenerating, but in some flowering plants it may contain two or all four of the tetrad nuclei (reviewed by Willemse and Van Went, 1984). The functional meiospore is described as a megaspore, and the sporangium that produces it as a megasporangium. Each megasporangium usually produces only a single megaspore in gymnosperms and flowering plants, and the megagametophyte (which is also called the embryo sac in flowering plants) that develops from this usually matures only a single embryo (although more than one may be initiated, especially in gymnosperms). In gymnosperms and flowering plants, the megaspore, and subsequently the megagametophyte, are retained within the megasporangium, and the entire complex, together with some additional sporophytic tissues (the integuments), is described as a seed. It is the seed, and not the megaspore, that disperses from the maternal sporophyte. Overwhelmingly, seed dispersal is delayed until the embryo has matured, although in a few gymnosperms (e.g. *Ginkgo biloba*) seeds are shed at about the time of fertilization, which can occur on the tree or after the seed has dispersed (Eames, 1955). This situation can be contrasted with that in heterosporous pteridophytes where, with the exception of *Selaginella rupestris* (Sporne, 1975) the megaspore, or the megagametophyte developing from it, is released from the megasporangium; hence these are not seed plants.

The gametophytes of heterosporous taxa are usually smaller than those of homosporous taxa. Maximum reduction is seen in the microgametophytes (pollen) of gymnosperms and flowering plants, which can comprise only three nuclei at maturity, two of which are gametes. The megagametophytes (embryo sacs) of flowering plants contain a few more nuclei (most commonly eight, Fig. 2.4), but in most taxa they each mature only a single functional egg. Those of gymnosperms are larger and may produce several archegonia and hence several eggs.

Most tracheophytes are cosexual (i.e. the individual produces both male and female gametes), although some seed plants have achieved unisexuality (dioecy). However, all bryophytes are cosexual (according to the definition of the individual proposed in Chapter 1, which identifies the sporophyte and the gametophytes it gives rise to as parts of the ontogeny of an individual). Most are also homosporous, and in many of these cosexuality is achieved by each gametophyte differentiating both eggs and male gametes. That is, each gametophyte is cosexual. However, there are a number of homosporous bryophytes that do not achieve cosexuality in this way but through each individual producing a genetically determined and thus predictable mixture of male gametophytes and female gametophytes. In this latter group, there are no obvious differences in the size, or in any other aspect of the appearance, of meiospores that will give rise to male gametophytes and those that will give rise to female gametophytes. However, in addition to homospory, anisospory is also found in the bryophytes. Heterospory has not been recorded. Anisosporous forms produce male gametophytes from microspores and female gametophytes from megaspores. Anisospory in bryophytes differs from heterospory in tracheophytes in that each meiosis in an anisosporous bryophyte produces a mixture (usually two of each) of microspores and megaspores (Vitt, 1968; Mogensen, 1983; Pant and Singh, 1989), whereas in a heterosporous tracheophyte there are separate male and female meioses. The microgametophytes produced by anisosporous bryophytes may be small, and in some taxa are epiphytic on the megagametophytes, but the megagametophytes remain large. The retention of a large size by the megagametophytes reflects the dominant role the gametophyte plays in the bryophyte life cycle.

There is some debate over the terminology that should be used to describe the gender state of bryophyte gametophytes (Wyatt and Anderson, 1984; Zander, 1984; Wyatt, 1985). The problem has arisen because the terminology usually applied to tracheophytes (e.g. dioecious, monoecious etc.) describes the capacity of the sporophyte to give rise, via its gametophytes, to gametes of different gender. Thus a dioecious tracheophyte produces either male gametophytes (microgametophytes) or female gametophytes (megagametophytes) but not both. A monoecious tracheophyte produces both. However, a tracheophyte may achieve monoecy either by producing a mixture of male gametophytes and female gametophytes (e.g. heterosporous individuals) or by producing cosexual gametophytes (e.g. homosporous individuals). The potential for confusion if the same descriptive terminology is applied to the sporophyte and gametophyte is plain: a dioecious individual produces dioecious gametophytes but a monoecious individual may produce dioecious or monoecious gametophytes. Because the gametophyte is the dominant stage in the bryophyte lifecycle, some bryologists have been keen to avoid this kind of confusion. It has been proposed that confusion can be avoided by changing the ' –oecious' endings used to describe the gender state of tracheophyte sporophytes to ' –oicous' endings (Zander, 1984). Thus a gametophyte that produces only one type of gamete is dioicous, whereas a cosexual gametophyte is monoicous. Zander's proposal is rightly gaining support among many bryologists (e.g. Mishler, 1988). I would argue that it would also be worthwhile to adopt it to describe the gender state of tracheophyte gametophytes. Thus it can be stated that all bryophytes are cosexual (monoecious), but an individual may achieve this condition either by producing a mixture of male and female gameto-

phytes – dioicous gametophytes – or by producing monoicous gametophytes. In contrast, gametophytes produced by seed plants are dioicous. However, an individual may generate a mixture of male and
female gametophytes (pollen and embryo sacs), in which case it is
cosexual (monoecious or hermaphrodite), or it may generate only
male, or only female, gametophytes, in which case the individual is
unisexual – or dioecious. Thus 'dioecy' describes an individual's state,
'dioicy' the state of its gametophyte parts: all seed plants are dioicous,
but only some are dioecious; no bryophyte is dioecious, but many are
dioicous.

There are a number of other aspects of sexual reproduction that
need to be considered. However, discussion of these will be postponed
to later sections when their relevance will be more apparent. Attention
will now be shifted towards a description of the patterns of asexual
reproduction in bryophytes and tracheophytes.

2.4 SEXUAL AND ASEXUAL REPRODUCTION IN BRYOPHYTES AND TRACHEOPHYTES

There is a complex and often imprecise terminology associated with
asexual reproduction. Any terms used will be defined to prevent any
confusion over their exact meaning. The major terms used are parthenogenesis, apogamy, apospory and diplospory.

Parthenogenesis involves the development of an embryo (a new
sporophyte in plants) from an unfertilized egg. Apogamy involves the
production of an embryo from vegetative tissues of the gametophyte.
This is a mitotic process. Consequently, the apogamously generated
sporophyte will have the same number of chromosomes as the gametophyte that produces it. The gametophyte cell that develops apogamously and the egg that develops parthenogenetically are zygote equivalents. Thus apogamy and parthenogenesis qualify as reproductive
processes. They qualify as forms of asexual reproduction because the
initiating cell is not formed by a process that involves syngamy.

Apospory involves the production of a gametophyte from vegetative
tissues of the sporophyte. This is a mitotic process, and it follows that
the gametophyte will have the same number of chromosomes as the
sporophyte that produces it. Diplospory involves the production of a
gametophyte from a meiospore (megaspore in heterosporous taxa) that
has been produced by a process that does not involve meiotic
reduction. Meiosis may be restitutional or it may be replaced by an
essentially mitotic division; consequently, the gametophyte will have
the same number of chromosomes as the sporophyte that produces it.
However, the production of gametophytes is a process of growth, not

of reproduction. But in apomictic taxa the gametophytes produced aposporously and diplosporously differentiate eggs that develop parthenogenetically. The viability of progeny will be heavily dependent on them inheriting the maternal genome intact, a requirement that is met by the aposporous or diplosporous production of gametophytes. Thus in these taxa, apospory and diplospory can be viewed as being integral components of a process that culminates in successful asexual reproduction. However, it is important to maintain the distinction between the diplosporous and aposporous 'growth' of gametophytes and the production of progeny by asexual reproduction. I will shortly be describing how apospory in pteridophytes seems to be associated with zygote production, not with parthenogenesis. In these taxa, apospory causes an aberration of the sexual cycle of reproduction; it is not an integral component of an asexual cycle of reproduction.

Reproduction in bryophytes

Bryophytes, of which there are approximately 25 000 species, are either homosporous or anisosporous. Self-incompatibility has been recorded in some taxa, but it is thought to be relatively rare (Smith, 1978; Miles and Longton, 1990), although incompatibility barriers have been observed in *Physcomitrella patens* (Ashton and Cove, 1977; Courtice *et al.*, 1978; reported in Mishler, 1988). Sexual reproduction involving gamete fusion can therefore be by allogamy, autogamy or non-parthenogenetic automixis. However, ecological factors will affect the relative incidence of these processes. Bryophyte meiospores have evolved as efficient units of dispersal: in many taxa they comprise the major mechanism of gene flow, as eggs and embryos are not dispersed but develop *in situ* on the gametophyte, and male gametes (which must swim to the archegonium through a film of free water) can rarely be dispersed over more than a few centimetres and often over no more than a few millimetres (Richardson, 1981; Wyatt, 1982). For most taxa, the probability of two meiospores from the same meiosis landing close enough together to produce gametophytes that can cross-fertilize must be very remote. Thus intergametophytic non-parthenogenetic automixis will be rare. However, self-compatible cosexual gametophytes will be able to self-fertilize. Consequently, intragametophytic non-parthenogenetic automixis may be common, although whether this is the case is unknown (Newton, 1990) and doubts may be raised about whether the 100% homozygous individuals that would be generated by this process would be reproductively successful in the field. It is difficult to judge whether autogamy will be common or rare. This will be affected by population density – which will determine the level of mixing of meiospores from different individuals during dispersal, and consequently of the likelihood of gametophytes from the same individual cross-fertilizing – and by the proportion of meiospores that are locally dispersed. The incidence of allogamy will be affected by how thoroughly the gametophytes from different individuals are intermixed, and on their density. The latter is important, given the very restricted dispersal range of male gametes. At low densities, cross-fertilization between gametophytes may be precluded. This will lead to a failure to reproduce sexually if the gametophyte is dioicous, or to monoicous gametophytes having to rely on intragametophytic non-parthenogenetic automixis to achieve reproductive success.

Growth through vegetative propagation (e.g. through fragmentation or gemmae production) is ubiquitous within the bryophytes (Parihar, 1965; Watson, 1971). (Bryologists typically describe this process as one of asexual reproduction (e.g. Mishler, 1988; Miles and Longton, 1990; Newton, 1990) although it does not fall within the definition of asexual reproduction given in Chapter 1.) Indeed, some liverworts of temperate and arctic regions, and many mosses, appear to propagate themselves purely by vegetative means at least over parts of their range (Schuster, 1966; Newton, 1984; Reese, 1984). Typically, this involves a gametophyte propagating copies of itself. For example, approximately 18% of the British moss flora and 9% of the eastern North American moss flora are found only as gametophytes, and approximately 40% of the British moss flora produce the sporophyte phase of the life cycle only very occasionally (Longton and Miles, 1982; Lane, 1985; Miles and Longton, 1990). Asexual reproduction appears to be much less common. There are no obligately asexual bryophytes! (By this, I mean that no lineages produce embryos exclusively by asexual processes.) However, asexual processes can occur alongside sexual processes in some individuals. Thus, apogamy can be experimentally induced in a number of species that usually reproduce sexually. This is most common in mosses (Bopp, 1968; Lal, 1984). But reports of it occurring naturally are restricted to a tetraploid (sporophytically tetraploid) form of *Phascum cuspidatum* (Springer, 1935). The aposporous production of gametophytes can also be induced in the laboratory in numerous species of mosses and liverworts. The aposporously derived gametophytes are typically smaller, grow more slowly, and are less prone to initiate gamete production than gametophytes produced from meiospores (Lal, 1984). However, I have been unable to find any reference to aposporously produced gametophytes initiating sporophytes apogamously. Instead, they seem to initiate them, if at all, as zygotes. Because aposporously produced gametophytes are not meiotically reduced, the zygotes will have a higher ploidy level than their parents. Clearly, these instances of apospory cause an aberration to the sexual cycle of reproduction. Smith (1978) argues that apospory is unlikely to occur in nature, describing only one doubtful report of it occurring naturally. Wyatt and Anderson (1984) disagree, arguing that the burial of sporophytes following wounding or grazing should be reasonably common and could result in apospory. As evidence, they point out that Smith (1978) shows that polyploidy is more frequent in mosses than in liverworts (78.7% and 10.5% respectively). They suggest that this is because apospory is more common in mosses than in liverworts, as mosses have longer-lived, photosynthetic sporophytes (the sporophytes of liverworts generally lack chlorophyll) and are frequently found on unstable substrates where accidental burial of sporophytes is likely. Burial will favour apospory.

Diplospory has been observed in liverworts but appears to be comparatively rare in mosses. Interestingly, although there are frequent reports of meiotic irregularities in bryophytes, reports of complete asynapsis are rare (Smith, 1978).

The only report of parthenogenesis in bryophytes – either in nature or in culture – is provided by Lal (1984), who found it to be a common laboratory phenomenon in cultures of *Physcomitrium pyriforme*, *P. coorgense* and *P. cyathicarpum*. The parthenogenetically induced sporophytes were smaller than sexually derived ones. However, they produced viable meiospores.

Reproduction in pteridophytes

There are more than 12 000 species of pteridophytes. Most are homosporous but heterospory is found in eight extant genera (*Selaginella*, *Isoetes*, *Stylites*, *Marsilea*, *Pilularia*, *Regnellidium*, *Salvinia* and *Azolla*). The gametophytes of homosporous taxa each have the capacity to produce both male and female gametes (i.e. they are innately monoicous) although either the male, or much more commonly the female, function may not be expressed (Parihar, 1965; Klekowski, 1969, 1973; Lloyd, 1974; Atkinson, 1975; Cousens, 1975, 1988). Self-incompatibility does not appear to be widespread. Consequently, sexual reproduction involving gamete fusion can be by allogamy, autogamy and non-parthenogenetic automixis, although the relative frequencies of these will be constrained by ecological factors in the same way as they are in bryophytes (e.g. Soltis and Soltis, 1987). A factor that may reduce the incidence of intragametophytic non-parthenogenetic automixis (which is usually described as intragametophytic selfing in the pteridophyte literature, e.g. Cousens, 1988) is the production of antheridia and archegonia at different stages of gametophyte development. This phenomenon is well developed in some taxa (Naf, 1979). Antheridia are produced before archegonia. However, there is often an overlap in the production of these organs, providing

a window within which intragametophytic non-parthenogenetic automixis can occur. A factor that may promote intragametophytic non-
parthenogenetic automixis in some taxa is the production of subter-
ranean gametophytes (these have a symbiotic relationship with a
fungus), as in *Botrychium* (McCauley *et al.*, 1985; Soltis and Soltis,
1986). Thus, the extent to which intragametophytic non-parthenogenetic automixis does occur is in some doubt. Klekowski (1973, 1979)
argues that it may be common. But it is evident that some populations
maintain high levels of heterozygosity, indicating that even in the
absence of self-incompatibility systems, outcrossing may be favoured
by some species (Lloyd, 1974; Klekowski, 1979; Soltis and Soltis,
1987). Even if intragametophytic fusions between gametes is a regular
phenomenon, the 100% homozygous sporophytes that would result
may be weak or inviable. There is a growing body of evidence that
this is the case in some species, although it is by no means the case
in all (reviewed by Cousens, 1988).

Asexual reproduction is widespread, being more common among
pteridophytes than in any other class of tracheophytes. Apospory
occurs sporadically in, and can be induced throughout, the pteridophytes. Apogamy is also seen in a number of species. However, no pteridophyte appears regularly to reproduce by the aposporous production of
gametophytes followed by the apogamous production of sporophytes,
even though this system, being completely mitotic, would maintain
both heterozygosity and a constant chromosome number from generation to generation. Nor is apospory associated with parthenogenesis,
which is extremely rare in this group. Consequently, as in bryophytes,
apospory should be seen as causing an aberration to the sexual cycle
of reproduction, rather than being seen to be an integral component
of an asexual cycle of reproduction.

In contrast to apospory, apogamy is a key element of regular systems
of asexual reproduction in the ferns. Indeed, approximately 10% of
homosporous pteridophytes are asexual (or 'agamosporic', Walker,
1966, 1979; Love *et al.*, 1977). All reproduce embryos apogamously
(i.e. none are parthenogenetic) and, with the possible exception of
Matteuccia orientalis (Lovis, 1977), all are obligately asexual. Sporophytes are produced from gametophytes that are derived from meiospores that have retained the sporophytic chromosome number. This
juxtapositioning of apogamy with the effective avoidance of meiotic
reduction during meiospore production results in a system of asexual
reproduction in which the chromosome number is maintained constant from generation to generation.

By far the most common form of agamospory is described as the
'Normal' type by Walker (1966), as automixis by Mogie (1986a), and as
the Döpp–Manton scheme by Klekowski (1973). The latter synonym,

extended to the 'Döpp–Manton scheme of agamospory', has been adopted in this discussion (see Fig. 2.1 and associated discussion). It is pertinent to compare this asexual process with the normal sexual process. During sexual reproduction, an archesporial cell in a sporangium undergoes four successive mitoses, producing a total of 16 spore mother cells. Each of these undergoes a reductional meiosis, resulting in an archesporial cell giving rise to 64 meiotically reduced meiospores. In many taxa, only one archesporial cell per sporangium will give rise to meiospores (Walker, 1966) although more than one may do so in some taxa (Sporne, 1975). During the Döpp–Manton scheme of agamospory, each dividing archesporial cell also undergoes four successive mitoses. However, although the first three of these divisions are normal, the fourth is endomitotic. As a result, only eight spore mother cells are produced, each of which has double the number of chromosomes of the sporophyte. These undergo reductional meioses, resulting in a total of 32 meiospores per dividing archesporial cell. Each meiospore has the same number of chromosomes as the sporophyte, and the gametophytes developing from them give rise to sporophytes by apogamy (Döpp, 1932; Manton, 1950).

Another type of agamospory, the Braithwaite scheme (Braithwaite, 1964), is seen in a few species. Here, the four mitoses preceding meiosis are normal and 16 spore mother cells are produced by each dividing archesporial cell. However, the first division of meiosis is restitutional. The second division proceeds normally so that each meiosis produces a dyad of unreduced meiospores rather than a tetrad of reduced meiospores. The sporangium therefore obtains a total of 32 meiospores per dividing archesporial cell. The gametophytes developing from these give rise to sporophytes by apogamy. The differences between the Döpp–Manton scheme, the Braithwaite scheme and sexual reproduction are illustrated in Fig. 2.5.

Clearly, the 50% reduction in meiospore production that results from the transition from sexual reproduction to asexual reproduction by the Döpp–Manton or Braithwaite schemes heralds a potential 50% cost of asexuality when compared with sexuality (Mogie, 1990). This cost will be investigated in detail in Chapter 3. It will arise whenever the transition is not accompanied by a compensatory increase either in the number of archesporial cells that divide to give spore mother cells or in the number of mitoses that separate the archesporial cell from its spore mother cell derivatives. No such compensatory increases have been reported.

Before leaving this part of the discussion it is pertinent to point out that the gametophytes produced by obligately agamosporous ferns usually have functional antheridia but lack, or produce rudimentary or abortive, archegonia. That is, they are female-sterile. This helps to

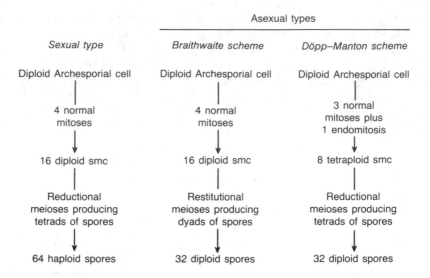

	Asexual types	
Sexual type	*Braithwaite scheme*	*Döpp–Manton scheme*
Diploid Archesporial cell	Diploid Archesporial cell	Diploid Archesporial cell
↓	↓	↓
4 normal mitoses	4 normal mitoses	3 normal mitoses plus 1 endomitosis
↓	↓	↓
16 diploid smc	16 diploid smc	8 tetraploid smc
↓	↓	↓
Reductional meioses producing tetrads of spores	Restitutional meioses producing dyads of spores	Reductional meioses producing tetrads of spores
↓	↓	↓
64 haploid spores	32 diploid spores	32 diploid spores

Figure 2.5   The pattern of spore production in sporangia of homosporous ferns reproducing either sexually, or asexually by the Braithwaite and Döpp–Manton schemes of apomixis. The numbers of spore mother cells (smc) and spores are those found in many taxa. However, some taxa may utilize more than one archesporial cell per sporangium in spore production. In these, the total spore production per sporangium will not be that depicted in the figure but the 50% reduction in spore production shown following the adoption of the Braithwaite or Döpp–Manton schemes will still be obtained.

explain why parthenogenesis is rare in this group. But more importantly, it provides the key to understanding why obligate asexuality should be more common in this group than in any other, despite its potential cost. This will be discussed in Chapter 3.

Reproduction in gymnosperms

There are approximately 730 gymnosperm species. Individuals are heterosporous, and often monoecious, seed plants, producing dioicous gametophytes. Consequently, sexual reproduction by gamete fusion can be by allogamy or autogamy. Sexual reproduction by non-parthenogenetic automixis has been observed as a very occasional phenomenon in several species; but this is a derived process that differs from non-parthenogenetic automixis in bryophytes and pteridophytes in that only one of the fusing nuclei is enclosed in a cell that is differentiated as a gamete. The gametic cell is always the egg. It is the male gamete whose role is usurped. For example, in some species the ventral canal cell of the archegonium can fuse with the egg nucleus. This has been recorded in *Encephalartos villosus* (Sedgewick, 1924), cycads

(Chamberlain, 1935), *Picea vulgaris, Pinus laricio, Abies balsamea* and *A. pindrow* (Dogra, 1966), although on the single occasion that it was recorded in *A. pindrow* the fusion was incomplete and no viable embryos were evident. Dogra (1966) interprets this form of automixis as asexual. It is more appropriate to classify it as an aberrant form of sexual reproduction as it involves the initiation of an embryo through nuclear fusion. The fact that the nucleus that fuses with the egg nucleus is not donated by a male gamete is interesting, but it comprises an aberration rather than a renunciation of the sexual process. Perhaps the most important fact to emerge from these reports of automixis is that cells in the axial row of the archegonium other than the egg cell can fulfil the role of a gamete. Emphasis will be placed on this in the next chapter, when the evolution of the female gametophyte of flowering plants (the embryo sac) is discussed.

Obligate asexuality is absent from the gymnosperms! So is apogamy. However, processes associated with asexual reproduction are widely distributed, although their presence appears to have minimal impact on the genetic structure of populations. Thus some individuals may occasionally produce offspring parthenogenetically. This has been reported in *Pseudotsuga menziesii* (Allen, 1942; Orr-Ewing, 1957), and in *Pinus wallichiana* and *P. nigra* var *austriaca*, but in *Pinus*, at least, the developing eggs were haploid and the resulting embryos failed to develop to maturity (Dogra, 1966).

Parthenogenesis has minimal impact on the genetic structure of populations because it appears to be both rare and rarely successful. However, another form of asexual reproduction that also has minimal impact is cleavage polyembryony, which is both widespread and successful (Foster and Gifford, 1959; Sporne, 1965; Willson and Burley, 1983 give detailed accounts of this phenomenon). Here, a (sexually produced) embryo develops from a zygote; but early in its development some of its cells appear to obtain the status of zygote equivalent and each divides to give rise (asexually) to an embryo. Consequently, a developing seed can contain numerous embryos derived (via zygote equivalents) from the same zygote. For example, the zygote of *Pinus* divides to give a 16-celled proembryo arranged in four superimposed tiers of four cells. Eight of these cells, including the four that comprise the bottom tier, typically give rise to embryos. As a megagametophyte of *Pinus* may differentiate six archegonia, and as each of the sexually produced zygotes developing from these may give rise to eight embryos by cleavage polyembryony, up to 48 embryos may be initiated during the development of a seed. However, as is generally the case with polyembryony in gymnosperms, all but one embryo will die as the seed matures. In *Pinus*, the surviving embryo is usually one of the embryos that have developed from a cell in the bottom tier of cells of a zygotically derived embryo (Foster and Gifford, 1959). Although this has been produced asexually it inherits the genome of the sexually produced parental embryo. Following germination and growth to maturity it will reproduce zygotes sexually. These will give rise, asexually, to embryos by cleavage polyembryony, one of which will survive at seed maturity. And so on. From a population genetics viewpoint, a population exhibiting cleavage polyembryony can be treated as though it is as an obligately sexual one reproducing by allogamy or autogamy!

Reproduction in angiosperms

There are approximately 250 000 species of flowering plants. Individuals are heterosporous, and often cosexual (monoecious or hermaphroditic), seed plants, producing dioicous gametophytes. Many are self-compatible. Consequently, sexual reproduction by gamete fusion can be by allogamy or autogamy. Sexual reproduction by parthenogenetic automixis may occur occasionally in *Brassica*. Here, an embryo is formed from the fusion of embryo sac nuclei (Eenink, 1974). This book is not the place to provide a review of the great wealth of reproductive processes and strategies found among sexual taxa. Detailed accounts are provided by Richards (1986) and in Lovett Doust and Lovett Doust (1988). Reviews of the development of reproductive organs and of the events associated with fertilization and embryo growth are provided in Johri (1984).

Several forms of asexual reproduction are found in this class. The most common are generative (diplosporous) and aposporous apomixis and polyembryony. A few species reproduce by the *Allium* scheme of parthenogenesis, and by replicative haploid parthenogenesis.

Polyembryony has been reviewed by Lakshmanan and Ambegaokar (1984). It occurs as an occasional phenomenon in many taxa, including *Lilium* and *Nicotiana* (Cooper, 1943) and *Aristolochia* (Johri and Bhatnagar, 1955), and as a regular phenomenon in others. It takes several forms. For example, several cells of the embryo sac, in addition to the egg cell, may initiate embryos (Kimber and Riley, 1963). This is a form of apogamy. However, these embryos do not replace the sexually derived embryo but co-exist with it in a multi-embryo seed. Embryos may also be produced from cells in the ovule outside the embryo sac (adventitious embryony). Richards (1986) reviews this process. It is a regular phenomenon in many taxa, and it is usually to be found in taxa that are subtropical or tropical, woody, diploid, and with fleshy, multiseeded fruits. For example, it is common in *Citrus* (oranges, lemons, grapefruits etc.), *Mangifera* (mango), *Euonymus* (spindles) and *Capparis* (capers). Usually, one of the asexually produced embryos invades the embryo sac, outcompeting both the other asexual embryos and the sexually produced embryo, and usurping the sexually produced endosperm. In some cases, however, the sexual embryo, and one or more of the asexual embryos co-exist, sharing the same (sexually produced) endosperm. Embryos may also be derived from cells of the sexually produced embryo (cleavage polyembryony). This process has been observed in *Erythronium americanum*, *E. dens-canis*, *Tulipa gesneriana*, *Cocos nucifera*, *Primula auricula*, *Corydalis cava*, *Eulophia epidendraea*, *Habenaria platyphylla*, *Geodorum densiflorum*, *Cymbidium bicolor*, *Stachyurus chinensis* and *Hamamelis virginiana* (Jeffreys, 1895; Swamy, 1942, 1943, 1949; Whitehead and Chapman, 1962; Mathew and Chaphekar, 1977; Mathew, 1980; Lakshmanan and Ambegaokar, 1984).

The *Allium* scheme of parthenogenesis has been recorded in *Allium nutans* and *A. odorum* (Hakansson, 1951; Hakansson and Levan, 1957). It is similar to the Döpp-Manton scheme of agamospermy in ferns, except that the embryo is formed parthenogenetically instead of apogamously. Thus the megaspore mother cell is formed by an endomitotic division and, as a result, has twice as many chromosomes as the maternal sporophyte. It undergoes a reductional meiosis, generating a megaspore with the sporophytic number of chromosomes. The egg of the embryo sac derived from the megaspore develops parthenogenetically.

Parthenogenetic automixis has been observed in *Rubus nitidoides* (Thomas, 1940) and replicative haploid parthenogenesis has been observed as an occasional phenomenon in *Brassica*. Here, the embryo is initiated by the parthenogenetic development of a haploid egg, but the developing embryo achieves diploidy by the fusion of cleavage division nuclei (Eenink, 1974).

As stated, the most widespread, and the most intensively investigated, forms of asexual reproduction in flowering plants are generative apomixis (diplospory) and aposporous apomixis. Because the control of these two processes will be investigated in some detail in Chapter 5, it is worth spending some time on their description here.

Aspects of generative and aposporous apomixis in angiosperms

Generative and aposporous apomixis share a number of features in common. They both involve the parthenogenetic development of eggs, they each have a wide though uneven taxonomic distribution, affected individuals may be either facultatively or obligately apomictic and may be pseudogamous or non-pseudogamous, and both types of apomict are usually polyploid descendants of perennials which were typically cosexual and often diploid, herbaceous and outbreeding. Both types are more likely to be found among temperate, subarctic and arctic-alpine floras, being poorly represented in tropical and subtropical floras. The few exceptions to this format include a number of apomictic forms of *Antennaria*, which are descended from dioecious ancestors, and apomictic members of the *Potentilla argentea*, *Hieracium umbellatum*, *Ranunculus auricomus* and *Arabis holboellii* complexes, some of which are diploid.

The main feature over which generative and aposporous apomicts differ is in the derivation of the embryo sac (female or megagametophyte). The embryo sac of generative apomicts is derived from the generative cell (the megaspore mother cell, also described as the embryo sac mother cell). This is the same cell that gave rise to the embryo sac in their sexual ancestors. By utilizing this cell, generative apomixis involves a direct hijack of the female part of the sexual process. Embryo sac formation during sexual reproduction commences with the megaspore mother cell undergoing the two divisions of a reductional meiosis. In the majority of species, including the ancestors of most generative apomicts, three of the four meiotic tetrad cells degenerate, and the surviving member (the megaspore) gives rise to an eight-nucleate embryo sac by three cycles of mitosis. One of the eight nuclei differentiates as an egg. This pattern of embryo sac formation is described as the '*Polygonum*' type (*Polygonum* being the taxon in which it was first described) and is illustrated in Fig. 2.4. It is 'monosporic', as only a single member of the meiotic tetrad is involved in its development. Two main types of embryo sac development in generative apomicts are derived from the *Polygonum* type. These, named after the organisms in which they were first described, are the *Antennaria* and *Taraxacum* types. The *Antennaria* type is exhibited by most generative apomicts. Here, meiosis is omitted. Instead, the megaspore mother cell functions directly as a megaspore, and enters into the three mitotic cycles involved in *Polygonum*-type embryo sac formation. The *Taraxacum* type is exhibited by a number of apomicts within the family Asteraceae (Compositae), but is rare in other families. Here, the megaspore mother cell undergoes meiosis. The first division, however, is characterized by an attenuated prophase and is restitutional and largely or wholly asyndetic, with the result that the product of meiosis is an unreduced dyad. One of the dyad cells degenerates. The other functions as the megaspore, producing an eight-nucleate *Polygonum*-type embryo sac by three cycles of mitosis. The sexual ancestors of a few generative apomicts exhibited a tetrasporic, rather than a monosporic, pattern of embryo sac production. In such forms, each nucleus of the meiotic tetrad contributes to the embryo sac, undergoing one to three cycles of mitosis. The generative apomicts descended from this type of sexual ancestor undergo a pattern of embryo sac formation described as the *Ixeris* type. Here, the megaspore mother cell undergoes a division of uncertain status. Richards (1986) describes it as mitotic, but Nogler (1984) describes it as being essentially meiotic. According to Nogler, the prophase of this division is similar to that observed in the *Taraxacum* type, but anaphase corresponds to Anaphase II of meiosis, with chromatids separating. The two nuclei formed by this division remained enclosed within the same cell membrane, and each undergoes two cycles of mitosis to give an eight-nucleate embryo sac. The embryo of *Antennaria-*, *Taraxacum-* and *Ixeris*-type apomicts is initiated by the parthenogenetic development of the egg.

Whereas the embryo sac of generative apomicts is derived from a generative cell, that of aposporous apomicts is derived from a somatic cell of the ovule. This cell is usually a cell of the nucellus, and is typically located close to the chalazal pole of either the megaspore mother cell or of its dyad or megaspore. Two types of aposporous embryo sac development can be recognized – the *Hieracium* and *Panicum* types. The *Hieracium* type is widely distributed. Here, the somatic cell that will form the aposporous embryo sac expands and vacuolates, and its nucleus enlarges. Neighbouring somatic cells are typically compressed and resorbed during this process. Following these changes, the aposporous initial divides mitotically to form an essentially eight-nucleate embryo sac, although antipodal cells may proliferate in some taxa. The *Panicum* type is widely distributed within, but is restricted to, the Panicoideae and Andropogoneae of the Poaceae (Gramineae). Here, the aposporous initial divides mitotically only twice, to form a four-nucleate embryo sac which usually comprises two synergids, an egg and a single polar nucleus; antipodals are absent. The embryo of both the *Hieracium* and the *Panicum* types of aposporous embryo sac is initiated by the parthenogenetic development of the egg. Thus, in contrast to the situation in bryophytes and pteridophytes (where aposporously derived gametophytes produce embryos sexually), apospory in flowering plants is an integral component of a system of asexual reproduction. But note that because the aposporous embryo sac is derived from a somatic cell, rather than from the megaspore mother cell, the latter is potentially in a position to initiate a sexual embryo sac. Thus apospory does not hijack the sexual system. It merely runs in parallel with it. Indeed, in most apospores the megaspore mother cell enters meiosis, but in most cases either it, or more often the dyad, megaspore or embryo sac derived from it, degenerates, enabling the aposporous embryo sac to take over the ovule. The aposporous embryo sac is usually initiated just before or during the degeneration of the megaspore mother cell or its products (Nogler, 1984). It is still unclear whether the degeneration of the megaspore mother cell or its products results from innate inviability, or from competition from the aposporous initial or its products. Nor is it known whether the megaspore mother cell meiosis is typically syndetic and reductional. However, Nogler (1984) considers that the megaspore mother cells of many apospores are probably incapable of embryo sac development. Nevertheless, the megaspore mother cells of some species are known to be capable of developing viable embryo sacs and embryos. These compete with their aposporous counterparts. So successful can this competition be that aposporous reproduction can be effectively precluded, the individual behaving as an obligate sexual. This is the case in *Malus domestica*, where embryo sacs derived from megaspore mother cells usually prevail against those derived aposporously (Kryolova, 1976; Nogler, 1984).

The asexual production of embryos by aposporous apomixis will clearly be obligate in lineages that produce innately inviable megaspore mother cells, megaspores or megaspore-derived embryo sacs. But there is an important evolutionary implication of inviability. This concerns the initial selective advantage of apospory. Consider the two following scenarios, both of which are enacted in a cosexual, sexually reproducing population; this is the type of population in which apomixis is considered to have usually evolved. In both scenarios an individual acquires a dominant mutation that causes the megaspore mother cell or its products to be inviable. In the first scenario this is all that the mutation achieves and the mutant and its descendants (it is still male-fertile) that inherit a copy of the gene will have a much lower fitness than female-fertile members of the population. This lineage will persist only if it can re-acquire an effective female function. It does this by acquiring a mutation for apospory. In the second scenario the mutation causing the degeneration of the megaspore mother cell (or its products) also causes aposporous embryo sacs to be initiated. The linkage of these two phenomena would result if metabolites released during degeneration stimulated a neighbouring somatic cell to differentiate as an embryo sac. The egg of the embryo sac develops parthenogenetically. There is a fundamental difference between these scenarios: aposporous apomixis is initially selected as an escape from female sterility in the first but as an alternative to sexual reproduction via the female function in the second. In both, the aposporous mutant will emerge into an otherwise sexual population and its persistence will be determined by how well it operates in this environment. But it is important to establish which of the two scenarios has been most common if we are to understand the primary selective forces responsible for the evolution of asexual reproduction.

In my view, an immediate transition from sexual to aposporous embryo production has been more common than a transition that involves a period of female sterility. Genetic arguments can be mustered in support of this view and will be given in Chapter 5. But even in the absence of these a case can be made by drawing attention to the phenomenon of facultative apomixis, which is common among apospores. A facultatively apomictic apospore produces embryo sacs aposporously in some ovules but from meiotically reduced megaspores in others. The eggs of the former develop parthenogenetically whereas those of the latter are fertilized. It is unlikely that facultative apomicts have evolved from female-sterile sexual ancestors by the acquisition of a mutation for apospory and another for the re-instatement of the sexual female function. It is much more likely that they have evolved directly from female-fertile sexual ancestors by the acquisition of a mutation for apospory that achieves phenotypic expression in only some ovules. I can think of no reason why a gene for obligate apospory could not arise in the same way.

Facultative apomixis is also found among generative apomicts. Here, a proportion of the megaspores are produced by a reductional meiosis and the remainder by one of the processes described above that avoid meiotic reduction. The eggs derived from the former are fertilized while those derived from the latter develop parthenogenetically. Facultative asexual reproduction as a regular means of producing embryos is peculiar to the flowering plants; it has earlier been described how asexuality in the pteridophytes (the only other class in which the asexual production of embryos is found as a regular method of reproduction) is typically obligate. An explanation for this will emerge in Chapter 3, but it is clear that the acquisition of apomixis by flowering plants need not be accompanied by a total rejection of sexual reproduction via the female function. Even when this does occur it should be remembered that many affected cosexual lineages will still be able to reproduce sexually via the male function if they are within pollination distance of compatible sexual individuals. An effective male function has been observed in apomicts of several genera, including *Taraxacum* (Richards, 1973, 1986) and *Ranunculus* (Nogler, 1984) and has formed the basis of numerous investigations into the genetic control of apomixis, including those discussed in Chapter 5.

There is another major difference between some apomictic flowering plants and asexual individuals of other major taxonomic groups, namely that, even when the transition from sexual to asexual reproduction may be complete, reproductive success via the female function may still be totally reliant on females having access to mates. Affected individuals exhibit 'pseudogamous' forms of generative and aposporous apomixis. They require the presence of pollen if viable embryos are to be produced asexually.

Pseudogamy is widespread, though by no means universal. For example, it is common among apomicts in the Poaceae (with the exception of *Calamagrostis, Cortaderia jubata* and *Lamprothyrsus*) and in the Rosaceae (with the exception of *Alchemilla* and *Aphanes*), and among apomicts in *Ranunculus*, and *Hypericum*; but, with the exception of *Parthenium*, it is absent from the Asteraceae (Grant, 1981; Nogler, 1984). Apomicts which are non-pseudogamous are described as 'autonomous'. With the possible exception of *Nardus stricta* (Rychlewski, 1961) pseudogamy, when it occurs, appears to be obligate.

The dependency of pseudogamous apomicts on pollination results from the nature of the nutritional regime within which embryos achieve maturity in flowering plants. Basically, the flowering plant embryo is directly nurtured not by the maternal sporophyte, nor by the quasi-maternal female gametophyte (as it is in all other classes), but by the endosperm. This tissue is itself a product of fertilization, being formed by the fusion of (usually) two embryo sac nuclei (the polar nuclei of Fig. 2.4) and a male gamete nucleus. The endosperm is the mitotic derivative of this triple fusion product. Primary among its roles is the supply of nutritional and growth regulatory substances to the developing embryo. In the absence of an effective endosperm, the embryo will degenerate before it reaches maturity. To all intents and purposes, the endosperm can be viewed as a sterile worker (analogous to the sterile castes in social insects), which ensures that copies of its genes are passed on to the next generation by helping its reproductively competent sib – the embryo – to survive the early part of its journey towards reproductive maturity (Queller, 1983, and Willson and Burley, 1983, give contrasting accounts of the evolution of the endosperm and of double fertilization in flowering plants). The dependency of the embryo on the endosperm is absolute for most taxa (exceptions are discussed by Queller, 1983). This dependency imposes a considerable barrier to the evolution of asexuality in flowering plants, for not only does the evolution of asexuality require the avoidance of meiotic reduction and the induction of parthenogenesis – events that would be sufficient to generate successful asexual taxa in other groups – but it also requires that the avoidance of fertilization does not result in the production of an inadequate endosperm. In non-pseudogamous taxa, the fusion of the polar nuclei is sufficient to initiate an adequate endosperm. In pseudogamous taxa this does not appear to be the case and these have had to retain fertilization in a modified form. Following pollination of pseudogamous apomicts, the two male nuclei are deposited in the embryo sac. One fuses with the polar nuclei, initiating an adequate endosperm. The other is free to fertilize the egg – an event that must obviously be prevented. In most pseudogamous taxa, fertilization of the egg is avoided by the phenomenon of precocious embryony. Here, the egg divides parthenogenetically before pollination. By the time the male gametes enter the embryo sac the egg has developed into a several-celled embryo. In other pseudogamous apomicts the egg is still present when male gametes enter the embryo sac. The avoidance of fertilization in these taxa is discussed in Chapters 3 and 5.

The reliance by the asexual female function of pseudogamous apomicts on an active sexual male function has numerous implications. For example, the pollen donor, if it is unrelated to the recipient, gains no benefits from helping the recipient to produce viable embryos, as the donor's genes are incorporated into the sterile endosperm rather than into the reproductively competent embryo. Indeed, the donor may suffer costs. For example, the donor and recipient will often be cosexual apomictic neighbours, and the donor's own asexually produced progeny may have to compete against those of the recipient; thus by its actions in helping the recipient to reproduce, the donor is reducing the probability of its own offspring surviving to reproduce. This will be investigated in detail in Chapter 4.

Before concluding this chapter it is worth considering the incidence and distribution of generative and aposporous apomixis in flowering plants. Both have wide though uneven distributions. This can be seen from Table 2.3. Such apomicts are to be found in eight of the 10 superorders, in 26% of orders, in approximately 7% of families, and in approximately 0.9% of genera. The proportion of apomictic species is difficult to determine accurately, due to many taxa being split into numerous 'agamospecies'; the affinity between agamospecies and sexual species as taxonomic units is unclear (Richards, 1986, p. 427). For example, *Rubus* and *Hieracium* are considered to contain well over 2000 agamospecies each, and *Taraxacum* is considered to contain almost 2000 agamospecies. However, the *Taraxacum* agamospecies, for example, are grouped into approximately 25 sections, and it is debatable whether the section in apomictic taxa is closer in taxonomic status to a species than is an agamospecies. Bearing this in mind, and using a 'lumped' rather than a 'split' taxonomy to describe apomicts, it can be tentatively suggested that approximately 0.1% of species are either generative or aposporous apomicts. Most of these are contained within three families: the Asteraceae, Poaceae and, to a lesser extent, the Rosaceae.

It is pertinent to attempt to establish from Table 2.3 the number of independent origins of generative and aposporous apomixis within the angiosperms. Arriving at an estimate is made difficult on two counts. The first once again concerns the taxonomic treatment of apomicts. The tendency to split apomicts into numerous agamospecies may have resulted in individual apomictic lineages being allocated to more than a single agamospecies. The second concerns the capacity of apomicts

Table 2.3 The taxonomic distribution of aposporous and generative apomicts within the angiosperms. The division of the class into taxonomic units follows the classification system of Heywood (1978). The list of apomictic taxa is taken from Nygren (1967)

Superorder	Number of				Orders containing aposporous and generative apomicts	Families containing aposporous and generative apomicts	Genera containing aposporous and generative apomicts	Number of species
	Orders	Families	Genera	Species				
Magnoliidae	9	32	457	13 000	Ranunculales Laurales Casuarinales Piperales	Ranunculaceae Calycanthaceae Casuarinaceae Saururaceae	Ranunculus (A) Calycanthus (G) Casuarina (G) Houttuynia (G)	2 3 1 1
Hamamelidae	7	11	46	1 400	–	–	–	–
Caryophyllidae	4	13	595	11 000	Polygonales Caryophyllales Plumbaginales	Polygonaceae Amaranthaceae Plumbaginaceae	Atraphaxis (A) Aerva (G) Statice (G)	1 1 1
Dilleniidae	12	60	1000	25 500	Theales Capparales Urticales	Guttiferae Cruciferae Urticaceae	Hypericum (A) Arabis (G) Elastoma (A,G)	1 1 2A + 4G
Rosidae	17	89	3500	57 700	Sapindales Polygalales Santalales Myrtales Rosales	Rutaceae Malpighiaceae Balanophoraceae Thymelaeaceae Rosaceae	Skimmia (G) Hiptage (A) Balanophora (G) Wikstroemia (G) Alchemilla (A) Cotoneaster (A) Malus (A) Potentilla (A,G,AG) Rubus (A,G) Sorbus (A)	2 1 2 1 split split 8 2+1+2 split 2

Subclass					Order	Family	Genus	Count
Asteridae	9	44	3640	62 800	Gentianales	Gentianaceae	Cotylanthera (G)	1
					Asterales	Asteraceae	Antennaria (G,AG)	11 + 1
							Arnica (G)	3
							Artemisia (AG)	1
							Centaurea (A)	1
							Chondrilla (G)	9
							Cichorium (A)	1
							Coreopsis (A)	1
							Crepis (A)	9
							Erigeron (G)	3
							Eupatorium (G)	1
							Hieracium subg. Euhieracium (G)	split
							Hieracium subg. Pilosella (A)	split
							Ixeris (G)	1
							Leontodon (A)	1
							Leontopodium (G)	1
							Parthenium (AG)	2
							Picris (A)	1
							Rudbeckia (G)	6
							Taraxacum (G)	split
Alismatidae	4	16	59	555	—	—		
Commelinidae	9	20	1086	19 700	Poales	Poaceae	Agropyron (A)	1
							Anthephora (A)	1
							Bothriochloa (A)	4
							Bouteloua (G)	1
							Brachiaria (A)	2
							Calamagrostis (G)	8
							Callipedium (A)	2
							Cenchrus (A)	2

Table 2.3 (continued)

Subclass					Order	Family	Genus	No.
							Chloris (A)	4
							Dichanthium (A)	4
							Echinochloa (A)	1
							Eragrostis (G)	1
							Heteropogon (A)	1
							Hierochloe (A)	1
							Hyparrhenia (A)	1
							Nardus (G)	1
							Panicum (A)	4
							Papsalum (A,G)	6
							Pennisetum (A)	8
							Poa (A,G)	8A + 5G
							Setaria (A)	2
							Themeda (A)	2
							Tricholaena (A)	1
							Tripsacum (G)	1
							Urochloa (A)	4
Aprecidae	4	5	342	5 700	Arales	Araceae	*Aglaomeria* (A)	1
Liliidae	2	16	1233	26 900	Liliales	Amaryllidaceae	*Cooperia* (G)	1
							Zephyranthes (G)	1
						Liliaceae	*Allium* (G)	1

A = aposporous apomict; G = generative apomict; AG = a species that reproduces by both aposporous and generative apomixis; split = species number difficult to determine because of the way the genus has been treated taxonomically (see text).

to reproduce sexually either through the male function or, in facultative apomicts, through the female function. The opportunities for sexual reproduction are such that within many genera numerous agamospecies may share a common apomictic ancestor. As a consequence, it would be naive to assume that the number of independent origins of apomixis approaches the number of taxonomically delimited apomictic species. However, a reasonable minimum estimate of the number of independent origins will approach the number of genera that contain apomictic taxa. There are approximately 80 such genera. The justification for suggesting this as a minimum estimate is that, although many apomicts are obviously hybrids, the overwhelming majority appear to be intrageneric hybrids. Intergeneric hybridization must either be very rare, or it must typically result in inviable hybrids. Thus, whereas genes for apomixis may spread within a genus, it is unlikely that they will spread between genera. A genus which contains apomicts is likely to have acquired apomixis *de novo*.

Chapter 3

The costs, benefits and constraints of asexual reproduction in plants

3.1 INTRODUCTION

It is clear that the taxonomic distribution of asexuality in plants is markedly uneven. An absolute requirement to reproduce embryos by asexual processes (obligate asexuality) is absent from the bryophytes and gymnosperms, but it is common in the pteridophytes and it can be observed in a number of angiosperm species. Facultative asexuality is found only within the angiosperms. The reasons for this distribution are not immediately obvious. Obligate asexual reproduction is not obviously associated with either homospory or heterospory, as each of these categories contains one class in which it is present and one class in which it is absent (Table 3.1). This, together with the occurrence of occasional and sporadic asexual reproduction in all classes, indicates that neither homospores nor heterospores are constrained by physiology, biochemistry or ontogeny to reproduce sexually. Perhaps the pattern of asexuality is primarily determined by ecological considerations. Certainly, the forms of asexual reproduction that are most common among the pteridophytes and angiosperms tend to preserve the maternal genotype. It could be that the absence of obligate or facultative asexual reproduction among bryophytes and gymnosperms is due to their having a greater requirement than pteridophytes or angiosperms to produce genetically variable progeny. This is unlikely. Vegetative propagation is widespread among bryophytes and is reasonably common among gymnosperms (Mogie and Hutchings, 1990), demonstrating that the propagation through time and space of a maternal genome is not inherently deleterious. Indeed, this method of propagat-

Table 3.1 A summary of the occurrence, mechanisms and selective value of asexual reproduction in bryophytes and tracheophytes

Taxon	Pattern of meiospore production	Obligate asexual reproduction	Facultative asexual reproduction	Most common methods of asexual reproduction	Selective value of asexual reproduction
Bryophytes	Homospory	Absent	Absent		
	Anisospory	Absent	Absent		
Pteridophytes	Homospory	Present	Absent[a]	Apogamy	Escape from sterility
	Heterospory	Absent	Absent		
Gymnosperms	Heterospory	Absent	Absent		
Angiosperms	Heterospory	Present	Present	Parthenogenesis	Alternative to sexual reproduction

[a] With the possible exception of *Matteuccia orientalis*.

ing genes is the only one available to several bryophyte species, which have lost their capacity for sexual reproduction (Chapter 2).

The types, as well as the incidence, of asexual reproduction are also taxonomically variable (Table 3.1). In homosporous plants, successful and regular asexual reproduction is always non-parthenogenetic and non-aposporous. Embryos are produced apogamously from gametophytes that are always derived from spore mother cells. In contrast, successful and regular asexual reproduction in heterosporous plants is predominantly parthenogenetic (it is always parthenogenetic when asexual reproduction is obligate), is frequently aposporous, but is never apogamous. In parthenogenetic forms, embryos arise from eggs produced by gametophytes that may be derived from spore mother cells (i.e. generative apomixis) but which may also be derived from somatic cells of the ovule (i.e. apospory). Heterosporous plants can also reproduce asexually by the production of embryos directly from sporophytic tissues – from somatic cells of the ovule in adventitious embryony, or from somatic cells of a sexually derived embryo in cleavage polyembryony. In homosporous plants, the equivalent process to adventitious embryony, which would comprise the production of embryos from vegetative cells of the capsule (in bryophytes) or sporangium (in tracheophytes) has not been recorded. However, a number of both heterosporous and homosporous taxa can generate small plants viviparously from non-reproductive organs of the sporophyte (e.g. on leaf margins).

Although these differences between heterosporous plants and homosporous plants are absolute with respect to regular and successful asexual reproduction, they do not necessarily reflect fundamental differences in the innate capacity of the two categories of organism to initiate particular patterns of asexual reproduction. Rather than this, they may simply reflect differences in the capacity of the two groups to benefit from a particular pattern. Thus, apospory is relatively common, and parthenogenesis is present but rare, in homosporous plants. And apogamy is present but rare in heterosporous plants (e.g. the development of embryos from synergids in flowering plants). However, these phenomena do not provide these organisms with regular, successful systems of reproduction. Parthenogenesis in homospores is not associated with the avoidance of meiotic reduction but results in haploid progeny. Apospory in homospores is not associated with apogamy or with parthenogenesis. Consequently, the meiotically unreduced gametophyte generated will behave sexually and will produce progeny that are at an enhanced ploidy level. This behaviour in an isoploid (2n, 4n, 6n etc.) population will generate anisoploid (3n, 5n, 7n etc.) progeny that have low fecundity. Similarly, apogamy in heterosporous seed plants is unlikely to be selectively advantageous because it simply generates competition within the seed between the embryo formed from the egg and the apogamously derived embryo. Thus for reasons of having to maintain a constant chromosome number or of avoiding competition, several of the asexual reproductive processes successfully adopted by some taxa cannot be utilized successfully by others, even though acquiring them may be within their capabilities.

Even if homosporous taxa could evolve a developmentally sustainable system of apospory (e.g. by simultaneously evolving apogamy in order to keep the chromosome number constant from generation to generation), it would be associated with considerable costs that are not associated with apospory in angiosperms. These arise because apospory in homosporous plants, unlike apospory in angiosperms, involves the manipulation of parts of the sporophyte that are not usually involved in sexual reproduction. A sporophyte of a sexual homosporous plant can produce, and disperse over large distances, many thousands or millions of gametophytes, as each gametophyte is initiated, and dispersed, as a minute meiospore. A sporophyte of an obligately aposporous mutant will be much less efficient at producing and dispersing its gametophytes, because a much larger volume of sporophytic tissue is required to produce a gametophyte aposporously than is required to produce a meiospore, and because aposporously derived gametophytes are not enclosed within a spore, which is the structure that has been honed by natural selection into an efficient unit of dispersal. Thus an aposporous mutant will produce fewer gametophytes, and hence fewer embryos, and will disperse these less efficiently than a sexual wild-type. These costs are not associated with apospory in angiosperms, as the aposporous gametophyte is produced from sporophytic tissues within the seed, and simply replaces the sexually produced gametophyte within this dispersal unit. Consequently, an aposporous mutant will produce the same number of gametophytes, and thus embryos, as a sexual wild-type, and will disperse its embryos just as effectively.

One final factor that must be taken into account if an attempt is to be made to understand the pattern of distribution of asexuality among bryophytes and tracheophytes is that asexual reproduction appears to have been favourably selected for different reasons in the two groups in which it is common. Among homosporous ferns, obligately asexual forms are female sterile. These either fail to initiate, or produce only rudimentary, archegonia. Lovis (1977) convincingly argues that this condition is probably a consequence of hybridization rather than of asexuality, the immediate ancestors of asexual lineages being female-sterile hybrids. Thus asexual reproduction among homosporous ferns appears to have been selected primarily as an escape from sterility. Sterility is certainly not a major reason for the adoption of asexual reproduction by angiosperms. It was argued in Chapter 2 that this possibility lacked any solid foundation with respect to apospory. It is even less tenable with respect to the evolution of generative apomixis as this pattern of reproduction is as dependent as the process of sexual reproduction on the megaspore mother cell and its products remaining viable. Thus, in contrast to the situation in ferns, asexual reproduction in angiosperms appears to have been selected primarily as an alternative to sexual reproduction.

The discord emerging from this survey, which is summarized in Table 3.1, has largely evaded analysis until now. In fact, it hides a simple melody, extracting the score of which is the main purpose of this chapter. Three principal players will emerge – ontogeny, phylogeny, and natural selection – which alternate the lead but which are ever present. The last of these will be given voice first, to illustrate a fundamental difference between homospores and heterospores, which is that there is a fundamental cost of sexual reproduction in heterosporous plants but of asexual reproduction in homosporous plants (Mogie, 1990). These costs largely explain the differences in the incidence of asexual reproduction between these groups. What differences remain can be explained by ontogeny. However, an awareness that the costs and benefits of sexual and asexual reproduction differ between homospores and heterospores is of no help when trying to understand why these groups have adopted different mechanisms of asexual reproduction. This understanding will emerge from a consideration of phylogeny and ontogeny. Nor is knowledge of the costs particularly helpful when the uneven distribution of asexuality within these groups is being investigated. Interpreting this requires that ontogeny is given centre stage. These last two issues will be investigated in turn once the first issue has been aired.

Throughout this chapter attention will be focused on the female function. However, it is important to remain aware that the overwhelming majority of asexual plants are cosexual and that many continue to express a viable male function. This function can play an important role in the spread and maintenance of asexual reproduction. This will become clear in the following discussion but the role of the male function will be considered in Chapter 4.

3.2 THE COSTS OF SEXUAL AND ASEXUAL REPRODUCTION IN HOMOSPORES AND HETEROSPORES

Maynard Smith (1971, 1978) demonstrates that some forms of sexual reproduction in anisogamous organisms are associated with a cost that is so great that, without any compensating factors, a mutation that simply switches reproduction from sexual to asexual would increase rapidly in frequency and eventually go to fixation. A description of this model provides an appropriate introduction to this section.

Consider a dioecious population that comprises equal numbers of sexual males (N_m) and sexual females (N_f) and a small number (n_A) of asexual females. The mutation for asexuality affects the pattern of reproduction, and results in progeny being identical to the mother, but it does not affect the capacity for reproduction or the fitness of progeny. Consequently, the number of eggs produced and of progeny successfully generated will be the same for both sexual and asexual females. Thus if each female produces k eggs, a proportion S of which contribute to successful progeny (whether by fertilization, in the sexual case, or by parthenogenesis, in the asexual case) then sexual females will successfully produce SkN progeny, and asexual females Skn. The cost of sex arises from the nature of the progeny, not from the numbers produced. Because the asexual female produces progeny that are simply copies of herself (that is, asexually derived progeny are female and asexual) the progeny generation includes Skn asexual females. However, sexual females must produce sons as well as daughters, and in this model they must produce them in equal numbers. Consequently, the SkN sexually derived members of the progeny generation comprise $0.5SkN$ males and $0.5SkN$ females. A sexual female produces only half as many daughters as an asexual female. As a result, the proportion of the population that comprises asexual females increases in one generation from

$$\frac{n_A}{N_m + N_f + n_A} \quad \text{to} \quad \frac{Skn_A}{0.5SkN_f + 0.5SkN_m + Skn_A}$$

or, as $N_m = N_f$, from

$$\frac{n}{2N + n} \quad \text{to} \quad \frac{n}{N + n}$$

When n is small relative to N, this approximates to a doubling in the frequency of asexual females each generation. Eventually, asexual females will replace sexual females (and males, as these are produced by sexual females). Thus there is a 50% cost of sex.

It is worth stressing that the cost of sex is not associated with fecundity. A sexual female produces the same number of offspring as an asexual female. The cost arises because the sexual female must produce sons as well as daughters. That is, the cost of sex is that of producing males.

Although the model clearly shows that there can be a considerable cost associated with sexual reproduction, the conditions under which this cost emerges are closely circumscribed. They include dioecy, equal fecundity of females, and a 1:1 sex ratio. Deviations from these conditions are common, and it is probable that one or other of them is violated by most, if not all, taxa. The question is whether violating these conditions fundamentally affects the message that when asexuality arises it will replace sexuality.

I will attempt to answer this specifically for anisogamous organisms, as this group includes the bryophytes and tracheophytes. The accepted view is that violations of the conditions present in Maynard Smith's model do not affect the fundamental message, with one exception – an obligately self-fertilizing organism will avoid the cost of sex. However, this view is mistaken. It arises because current models only consider a taxonomically biased sample of anisogamous organisms that share a particular method of generating eggs. This sample is admittedly large, comprising multicellular animals and heterosporous plants. But it is unrepresentative of anisogamous organisms as it does not include homosporous plants. This is a critical omission, as there is a fundamental cost of asexuality in these! Before I demonstrate this I wish first to examine violations that do not alter the message that there is a cost of sex. That is, I will deal first with the fate of asexual mutations in multicellular animals and heterosporous plants. This part of the examination draws on the work of Charlesworth (1980). Throughout, only dominant genes for asexuality will be considered. This greatly simplifies the analysis, but dominance is not assumed primarily for this reason. It is assumed because there is good reason to believe that genes for asexuality are often effectively dominant to those for sexuality, although dominance is not absolute and is due to dosage effects. The latter provisos need not concern us unduly here, and the whole issue of dominance will be fully investigated in Chapter 5.

One assumption of Maynard Smith's (1971) model is that a switch from sexual to asexual reproduction will not affect the number of eggs produced. This assumption is generally valid for multicellular animals and heterosporous plants. In sexual forms of both groups, a diploid generative cell (the oocyte in animals, the megaspore mother cell in heterosporous plants) undergoes a reductional meiosis to produce three or four haploid nuclei (depending on whether one or both of the products of the first division of meiosis are included in the second). With only a few exceptions (in some heterosporous plants), all but one of these nuclei degenerate, and the survivor becomes the nucleus of the egg (in animals) or of the megaspore (in plants). Thus each female meiosis produces a single egg or megaspore. Following a switch to asexual reproduction, the generative cell can act directly as a mega-spore (the *Antennaria* scheme) or it can undergo either a restitutional meiosis or a mitotic division. These patterns have already been described in some detail in Chapter 2. When it does not act directly as a megaspore, a dyad of unreduced nuclei is generated. One of these nuclei degenerates and the survivor becomes the nucleus of the egg or of the megaspore. Thus, although a switch to asexuality causes fundamental changes in the pattern of egg or megaspore production, it does not affect the number of eggs or megaspores produced by each meiosis. The assumption of the model that sexual and asexual forms produce the same number of eggs is therefore valid for these organisms. However, others of the assumptions of the model can reasonably be altered. For example, deviations from a 1:1 sex ratio are known to occur, and it is clear that females in a population will exhibit a range of fitnesses. Charlesworth (1980) investigates the consequences of these deviations in a model that sets the sex ratio at r, where r is the frequency of males and $(1 - r)$ the frequency of sexual females, and which assumes two fitness levels which are dependent on whether an individual is inbred or outbred.

A part a of the matings in the population are assumed to result in inbred progeny and a part $(1 - a)$ in outbred progeny. Progeny resulting from inbreeding have fitness $(1 - d)$ relative to a fitness of 1 for outbred progeny. Here, d is a measure of inbreeding depression ($0 \leqslant d \leqslant 1$). This makes it possible to follow the fate of a gene for asexuality that arises in an inbred individual with lower fitness, or an outbred individual with higher fitness. Charlesworth further assumes that the population is diploid, that it is initially fixed for an allele *A1* at a

particular locus, and that allele *A2* arises at this locus and is a dominant gene for asexuality. Sexual individuals are consequently *A1A1*.
Because the gene for asexuality is dominant, individuals which acquire
a copy will reproduce asexually. It is very unlikely that such individuals will acquire a second copy of this gene (a second copy could
only be acquired by a new mutation). All asexual individuals can
therefore be assumed to be heterozygous (*A1A2*) at this locus. Because
of this, the fate of the asexual mutation can be determined simply
by following how the proportion of females that are asexual in the
population changes between generations. In other words, it is only
necessary to determine the proportionate contribution to female offspring by the asexual component of the population. Let the proportion
of females that are asexual in a given generation be x, and let $(1 - x)$
be the proportion that is sexual. Let these proportions in the following
generation be x' and $(1 - x')$ respectively.

First, consider the fate of a gene for asexuality in an inbred female
with fitness $(1 - d)$. This will also be the fitness of her daughters, as
these are identical to her. The fitness of the progeny of sexual females
is affected by the proportion that are produced by inbreeding. There
are a such matings, producing progeny with a fitness of $(1 - d)$. $(1 - a)$ matings involve outbreeding, producing progeny with a fitness of
1. Thus the mean fitness of sexual females is $a(1 - d) + (1 - a) = 1 - ad$. The proportion of their progeny that is female is $(1 - r)$. Thus
the proportion of females in the progeny generation that is produced
by inbred asexual females is

$$x' = \frac{x(1 - d)}{x(1 - d) + (1 - x)(1 - r)(1 - ad)} \tag{3.1a}$$

The asexual genotype will spread to fixation if the number of daughters produced by an asexual female is greater than the mean number
produced by a sexual female. This is when

$$(1 - d) > (1 - r)(1 - ad)$$

An effective way to illustrate this is to simplify the denominator of
Equation (3.1a) by assuming that x is very small (which it will be
initially). Under this assumption $x(1 - d)$ is close in value to zero,
and $(1 - x)$ is close in value to 1. The denominator can be adjusted
accordingly, resulting in Equation (3.1a) being approximated by

$$x' = \frac{x(1 - d)}{(1 - r)(1 - ad)} \qquad (x \backsimeq 0) \tag{3.1b}$$

Dividing each side of this equation by x gives

$$\frac{x'}{x} = \frac{1 - d}{(1 - r)(1 - ad)} \qquad (x \rightleftharpoons 0) \qquad (3.1c)$$

The asexual genotype will increase in frequency when $x'/x > 1$, which is when $(1 - d) > (1 - r)(1 - ad)$.

Attention can now be turned to the situation in which an asexual mutation arises in an outbred female with a fitness of 1. This will also be the fitness of this mutant's daughters. This gives

$$x' = \frac{x}{x + (1 - x)(1 - r)(1 - ad)} \qquad (3.2a)$$

and

$$\frac{x'}{x} = \frac{1}{(1 - r)(1 - ad)} \qquad (x \rightleftharpoons 0) \qquad (3.2b)$$

Here, the asexual genotype will spread to fixation if $1 > (1 - r)(1 - ad)$. This will always be the case, as r, the frequency of males, must be greater than zero.

The net intensity of selection on an asexual mutant can be obtained by taking the weighted mean, $E(x'/x)$, of Equations (3.1c) and (3.2b). The weights are a and $(1 - a)$ respectively. When x is close in value to zero, this gives

$$E(x'/x) = \frac{a(1 - d)}{(1 - r)(1 - ad)} + \frac{1 - a}{(1 - r)(1 - ad)}$$

$$E(x'/x) = \frac{1 - ad}{(1 - r)(1 - ad)}$$

$$E(x'/x) = \frac{1}{1 - r} \qquad (x \rightleftharpoons 0) \qquad (3.3a)$$

Equation (3.3a) tells us that the initial advantage of an asexual mutation is determined only by the sex ratio, r. The equation is independent of the mating system of the population, and of the level of inbreeding depression (Charlesworth, 1980). Note that when $r = 0.5$ (which it frequently approximates to in dioecious species), the 50% cost of sex demonstrated by Maynard Smith (1971) is obtained. This cost increases when $r > 0.5$ and decreases when $r < 0.5$. In conclusion, violating the assumptions of Maynard Smith's model by allowing a deviation from a 1:1 sex ratio and by allowing females to exhibit differences in fitness does not change the fundamental message that when asexuality arises it will replace sexuality.

So far, the advantages and disadvantages of sexual and asexual reproduction have been investigated only for dioecious organisms. Most plants are cosexual. These will not experience a cost of producing males. Is there still a cost to sexual reproduction? In attempting to answer this question, it is very important to distinguish between heterosporous and homosporous plants. The heterosporous case will be dealt with first.

A heterosporous cosexual plant reproducing sexually undergoes two types of meiosis. One results in the production of microgametophytes (e.g. pollen) which differentiate male gametes, the other in the production of megagametophytes (e.g. embryo sacs) which differentiate eggs. As described in the previous chapter, each male meiosis produces four male gametophytes, but each female meiosis produced only a single female gametophyte (at least in seed plants; some heterosporous pteridophytes may produce more than one megagametophyte from each meiosis). A switch to asexuality does not fundamentally affect this. For example, generatively apomictic flowering plants still produce a single female gametophyte from each megaspore mother cell irrespective of whether the reductional female meiosis characteristic of sexual forms is replaced by a restitutional *Taraxacum*-type division or by a mitosised *Antennaria*-type division. Similarly, in aposporous apomicts the embryo sacs derived asexually from somatic cells in the ovule tend to replace the megaspore-derived embryo sacs on a one to one basis (several aposporous embryo sacs may be initiated in each ovule but only one tends to survive at seed maturity). As each embryo sac produces a single egg in both sexual and asexual forms, it follows that any analysis of the consequences of acquiring asexuality should be based on the premise that its acquisition does not alter the number of eggs produced, or their viability. Even so, it cannot necessarily be assumed from this that reproductive success via the female function will be the same in sexual and asexual individuals. For example, a shortage of pollinators may reduce the number of eggs fertilized in a sexual individual but it will not affect the reproductive success of autonomous apomicts.

The situation with respect to the viability of male gametes in sexual and asexual individuals may be somewhat different from that relating to the viability of female gametes. Perhaps surprisingly, many apomicts still undergo an essentially reductional male meiosis irrespective of the pattern of apomixis exhibited by the female function (Nogler, 1984). That is, the asexual mutation often seems to have little effect on the male function even though it has a major effect on the female function. This is very important, as it means that a cosexual asexual plant may be able to reproduce sexually via its male function even though it may be obligately asexual via its female function. Consequently, an asexual mutation may alter in frequency not only because of its effects on the female function, but also because it is being incorporated into sexually reproduced offspring via the male function. However, as most apomicts are polyploid, indeed often anisoploid, male meiosis will often be highly disturbed. Such disturbances may result in a total or partial failure of reduction in some divisions, resulting in pollen exhibiting a range of chromosome numbers, from the fully reduced number to the fully unreduced number (Nogler, 1984; Richards, 1973, 1986). Only a proportion of this pollen will produce viable male gametes that can contribute to viable embryos. Consequently, although there is no reason to assume that the transition from sexual to asexual reproduction will be accompanied by a change in the number or viability of eggs produced, there is reason to assume that, in some taxa, it may be accompanied by a reduction in the efficiency of the male function.

In the models that follow, the fate of a gene for asexuality will first be followed on the assumption that asexual and sexual individuals have equal success via their male function and via their female function. The models will then be modified to encompass the situation where this assumption does not hold. This is a reasonable approach as the assumption of equal success may be valid for some taxa, although it is clearly not valid for all. It also has the benefit of keeping the mathematics within manageable limits.

The fate of a gene for asexuality that arises in a cosexual heterosporous population can be examined by methods analogous to those used by Maynard Smith (1971) for the dioecious case. Male meiosis in the asexual mutants will be assumed to be reductional, and the probability of an egg producing a viable embryo will be assumed to be the same in sexual and asexual individuals. Once again, the gene for asexuality will be assumed to be dominant, so that asexual individuals are heterozygous for it and for the wild type allele. This means that the mutation is passed on to every progeny produced asexually via the female function (as the parental genome is replicated in each of its asexually produced progeny) but to only half the progeny produced sexually via the male function (as pollen is meiotically reduced). If it is assumed that each member of the population is equally competent as a male parent, then the proportion of eggs fertilized by asexual individuals is equal to the proportion of asexual individuals in the population. If the population comprises N sexual and n asexual individuals, this proportion is $n/(N + n)$.

Using k and S as before, to represent, respectively, the number of eggs produced by an individual and the proportion of these that produce successful progeny, then SkN progeny are produced sexually, and Skn asexually. However, $SkN[n/(N + n)]$ of the sexually produced progeny will have been sired by the asexual component of the population. Half of these progeny will have acquired the gene for asexuality. Thus in one generation, the proportion of asexual individuals in the population will change from

$$\frac{n}{N + n} \quad \text{to} \quad \frac{Skn + 0.5SkN[n/(N + n)]}{SkN - 0.5SkN[n/(N + n)] + Skn + 0.5SkN[n/(N + n)]}$$

or from

$$\frac{n}{N + n} \quad \text{to} \quad \frac{n + 0.5N[n/(N + n)]}{N + n} \tag{3.4a}$$

When n is small relative to N, which it will be initially, $(N + n)$ approximates to N and $0.5N[n/(N + n)]$ approximates to $0.5n$. Expression (3.4a) approximates to

$$\frac{n}{N + n} \quad \text{to} \quad \frac{1.5n}{N + n} \quad (n \simeq 0) \tag{3.4b}$$

Thus there is a 1.5-fold advantage to asexuality (or a 33% cost of sex) in cosexual heterosporous organisms. This is less than the 50% cost of sex determined for dioecious organisms by this approach, but it is still very considerable.

Expression (3.4b) is given under the assumption that an asexual individual is obligately so via its female function. This is often the case, but a number of cosexual apomictic flowering plants are facultatively apomictic, producing progeny sexually from meiotically reduced eggs and asexually from meiotically unreduced eggs. Expressions (3.4) can be suitably altered to accommodate this phenomenon. Let v be the proportion of progeny produced asexually via the female function, and $(1 - v)$ the proportion produced sexually. Thus $vSkn$ progeny are produced asexually and $(1 - v)Skn$ are produced sexually via the female function of facultative apomicts. Half the meiotically reduced eggs produced by these apomicts will receive a copy of the gene for asexuality and will give rise, on fertilization, to facultatively apomictic progeny. A proportion $n/(N + n)$ of the remaining reduced eggs will be pollinated by facultative apomicts, and half the resulting progeny will receive a copy of the gene for asexuality from the pollen. Note also that a proportion $n/(N + n)$ of the meiotically reduced eggs that contain the gene for asexuality will also be pollinated by facultative apomicts and half the resulting progeny will receive a copy of the gene for asexuality from the pollen. These will be homozyous for the gene for asexuality. These homozygotes make the calculations more complex, but initially, when n is very small relative to N, they can be safely ignored to the extent that an approximate measure of the fate of the gene for asexuality can still be obtained simply by estimating the change between generations in the proportion of the population that is facultatively apomictic. Thus n facultative apomicts produce

$$Skvn + 0.5Skn(1-v) + \frac{0.25Skn^2(1-v)}{SkN + Skn} = 0.5Skn(1+v) + \frac{0.25Skn^2(1-v)}{Sk(N+n)}$$

facultatively apomictic progeny and

$$0.5Skn(1-v) - \frac{0.25Skn^2(1-v)}{Sk(N+n)}$$

sexual progeny. The N sexual members of the population produce SkN progeny, but a proportion $n/(N + n)$ are pollinated by facultative apomicts and half of these receive a copy of the gene for asexuality. Thus (after cancelling out Sk throughout the right hand side of the equation) the change in the proportion of the population that is facultatively apomictic, during the early stage of the spread of the gene for asexuality is from

$$\frac{n}{N+n} \text{ to } \frac{0.5n(1+v) + (0.25n^2(1-v)/(N+n)) + (0.5nN/(N+n))}{\begin{array}{c} 0.5n(1+v) + (0.25n^2(1-v)/(N+n)) + (0.5nN/(N+n)) + \\ N + 0.5n(1-v) - (0.25n^2(1-v)/(N+n)) - (0.5nN/(N+n)) \end{array}}$$

Considerable rearrangement gives

$$\frac{n}{N+n} \text{ to } \frac{n(1+0.5v) - (0.25n^2(1+v)/(N+n))}{N+n}$$

This is based on the assumption that n is small relative to N. Second order terms in n can therefore be ignored, giving

$$\frac{n}{N+n} \text{ to } \frac{n(1 + 0.5v)}{N+n} \qquad (n \simeq 0) \qquad (3.4c)$$

Note that when $v = 1$, the individual is obligately apomictic via its female function, and expression (3.4c) is equal to (3.4b). In general, as $n < n(1 + 0.5v) < 1.5n$ (when $0 < v < 1$), the proportion of the population that is facultatively apomictic will increase, but the rate of increase will not be as rapid as that in a population where the gene for asexuality causes its carriers to reproduce by obligate apomixis via the female function. However, the rate in the facultatively apomictic case will increase as the proportion of apomicts that are homozygous for this gene increases, as all the pollen and all the meiotically reduced eggs produced by these will carry a copy of the gene for asexuality and will therefore contribute to asexual progeny.

To return to consideration of obligate asexual reproduction via the female function, it can be seen that Equation (3.4b) is given under the assumptions of random mating and equal fitnesses. Charlesworth (1980) has investigated the spread of a gene for asexuality in a cosexual heterosporous population under conditions in which the sexual component of the population can self-fertilize at frequencies other than those expected under random mating, and with inbred progeny suffering a reduction in fitness relative to outbred progeny. Here, a represents the fraction of zygotes produced by sexual individuals by selfing, with $(1 - a)$ representing the fraction produced by outbreeding. The fitness of the former is $(1 - d)$ relative to a fitness of 1 for the latter, with d being a measure of inbreeding depression $(0 \leq d \leq 1)$. As in the previous model, the gene for asexuality is assumed to be dominant, so that asexual individuals are heterozygous at the locus concerned, and the efficiency of the asexual component of the population as male parents is assumed to be the same as that of the sexual component. x is the frequency of asexual individuals in the population in one generation, and x' their frequency in the next.

Under this model, asexual individuals can be produced parthenogenetically via the female function of an asexual parent, or by the fertilization of an egg of a sexual individual with a male gamete provided by an asexual parent. Asexual individuals produced by the latter route are outbred and will have a fitness of 1. The fitness of asexual progeny produced parthenogenetically will equal that of their mother. If the mother is inbred they will have a fitness of $(1 - d)$. If she is outbred they will have a fitness of 1. We will first consider the case of a part of the asexual component of the parental generation being inbred. This situation will arise if the asexual mutation originally arises in an inbred individual. All the descendants of this individual produced via its female function will be inbred. Let the proportion of the asexual component of the population that is inbred by x_1, with x_2 being the proportion that is outbred (where $x = x_1 + x_2$).

The inbred asexual component of the population in the progeny generation is contributed solely by the female function of inbred asexual parents, as these parents produce outbred progeny via their male function. Consequently, these parents contribute $x_1 (1 - d)$ of inbred asexuals to the progeny generation. This can be interpreted as a measure of their fitness via their female function. The mean fitness of the parental population can be defined as the expected fitness in the absence of inbreeding minus the reduction in fitness due to inbreeding. The former has a value of 1. The reduction in fitness due to inbreeding can be measured as the product of d and the fraction of the population that is inbred. This fraction is made up of a component from the asexual part (x_1) and one from the sexual part $[a(1 - x)]$. Thus the mean fitness of the population is represented by

$$1 - d[x_1 + a(1 - x)] = 1 - x_1 d - (1 - x)ad$$

Consequently, the frequency of inbred asexual individuals in the progeny generation is

$$x_1' = \frac{x_1(1 - d)}{1 - x_1 d - (1 - x)ad} \tag{3.5a}$$

The progeny generation will also contain outbred asexual individuals. Some of these will be generated asexually via the female function of outbred asexual parents. Others will be generated sexually via the male function of both inbred and outbred asexual parents. As outbred individuals have a fitness of 1, the contribution to outbred asexual progeny by the female function of asexual parents is simply x_2. Determining the number of outbred asexual progeny produced sexually is slightly more complicated. The proportion of the population that is sexual is $(1 - x)$. Of the eggs that this proportion produces, a fraction $(1 - a)$ is available for fertilization by non-self male gametes. Of these, a proportion x will be fertilized by asexual individuals, but only half of this proportion will acquire a gene for asexuality. Thus, $0.5x(1 - x)(1 - a)$ asexual progeny will be produced sexually. These will have a fitness of 1. Consequently, the frequency of outbred asexual individuals in the progeny generation is

$$x_2' = \frac{x_2 + 0.5x(1 - x)(1 - a)}{1 - x_1 d - (1 - x)ad} \tag{3.5b}$$

Note that the denominator in Equations (3.5) is equal to or less than 1. It can be seen from Equation (3.5b) that the gene for asexuality will spread to fixation if $a < 1$. That is, the gene will spread to fixation if there is some outcrossing. It is worthwhile to deviate slightly and briefly from the main purpose of this section and consider the fate of a gene for asexuality which arises in a population of obligate selfers. Here, $a = 1$, $x_1 = x$ and $x_2 = 0$. Equation (3.5a) can be rewritten as

$$x' = \frac{x(1 - d)}{1 - xd - (1 - x)d}$$

or

$$x' = \frac{x(1 - d)}{1 - d} = x \tag{3.5c}$$

That is, under obligate selfing, the gene for asexual reproduction is selectively neutral. There is no cost of sex!

To return to the main theme, Equations (3.5a) and (3.5b) can be simplified if x_1 and x_2 are so small that $(1 - x)$ and $(x_1 d)$ can be neglected (as the former approximates to 1 and the latter to zero). Rewriting these equations after simplification gives

$$x'_1 = \frac{x_1(1 - d)}{1 - ad} \tag{3.5d}$$

$$x'_2 = \frac{x_2 + 0.5x(1 - a)}{1 - ad} \tag{3.5e}$$

If $a < 1$ then $(1 - ad) > (1 - d)$. Under this condition x_1 tends to zero. Thus the system can be adequately represented by setting $x_2 = x$ $(x_1 = 0)$. This gives

$$x' = \frac{x + 0.5x(1 - a)}{1 - ad} \tag{3.5f}$$

dividing throughout by x gives

$$\frac{x'}{x} = \frac{1.5 - 0.5a}{1 - ad} = \frac{3 - a}{2(1 - ad)} \tag{3.5g}$$

We can now turn attention to the situation of an asexual mutation first arising in an outbred individual. In this case, all asexual individuals have a fitness of 1, and $x_1 = 0$. Thus this system is represented by Equation (3.5b). When x is very small, Equation (3.5g) is again obtained.

In conclusion, it is clear that the initial selective advantage of a gene for asexual reproduction is independent of whether it arises in an inbred or an outbred individual. Note that with no inbreeding (i.e. $a = 0$) Equation (3.5g) gives,

$$\frac{x'}{x} = \frac{3}{2} \quad (x \rightarrow 0) \tag{3.5h}$$

illustrating the 1.5-fold advantage to asexuality obtained from the preceding model (Equation 3.4b).

To summarize the points made so far in this section, it can be shown that sexual reproduction in dioecious organisms, and in cosexual heterosporous plants that do not obligately self-fertilize, carries with it a cost such that a gene for obligate or facultative asexual reproduction will rapidly increase in frequency and will spread to fixation. However, this conclusion comes from models that have assumed that sexual and asexual individuals obtain equal reproductive success both as male parents and as female parents. But the point has already been made that this assumption may hold only for some taxa. It is therefore necessary to investigate the extent to which this assumption can be violated before the conclusion that asexuality will replace sexuality becomes invalid.

This problem can be approached by reintroducing Expression (3.4b) but in a slightly modified form. That is, when sexual and asexual individuals are equally efficient both as male and as female parents, the change between generations in the proportion of the population that is asexual is from

$$\frac{n}{N + n} \quad \text{to} \quad \frac{n + 0.5n}{N + n}$$

where N and n are, respectively, the number of sexual and asexual individuals in the parent generation. In this equation, the numerator of the right hand side is composed of two parts: that on the left describes the number of asexual progeny produced via the female function, that on the right the number produced via the male function. It is clear from this that, when reproductive success via the female function is the same in sexual and asexual individuals, then in the absence of an effective male function a gene for asexuality will be selectively neutral. It is equally clear that the male function needs to contribute to only a single successful progeny for asexuality to spread to fixation. Thus the fact that the male function of asexual individuals may often be much less effective than that of sexual individuals does not invalidate the general conclusion that an asexual mutation arising in a cosexual heterosporous population will be favourably selected. All that is required is that the male function is at least minimally effective.

The requirement for an asexual individual to be able to sire at least one progeny if asexuality is to replace sexuality exists only if reproductive success via the female function is not greater in asexual than in sexual individuals. Otherwise, a gene for asexuality will spread to fixation even in a male-sterile lineage. Conversely, if asexual individuals are less successful than sexual individuals as maternal parents a gene for asexuality will only increase in frequency if the male function can overcompensate for the shortfall (at a minimum, the male function must be able to make up the shortfall and contribute to one additional offspring). These considerations can be formalized by introducing into the equation the qualifying terms f and m, where f is the quotient of (fitness of the asexual component of the population via the female function/fitness of the sexual component of the population via the female function) and m is the equivalent quotient for the male function. That is, f and m are measures of the mean success of the female and male functions of asexual individuals relative to that of sexual individuals. Incorporating these into the modified form of Expression (3.4b) gives

$$\frac{n}{N+n} \quad \text{to} \quad \frac{fn + 0.5mn}{N + fn}$$

It follows that a gene for asexuality will increase in frequency when

$$f + 0.5m > 1 \text{ (approx.)}$$

Because male meiosis is often disturbed, the range of values of m will typically lie between 0 and 1. However, although asexual individuals undergoing generative or aposporous apomixis will produce the same number of eggs as conspecific sexuals they may be more successful at producing offspring via the female function. Under this circumstance, $f > 1$ and a gene for asexuality will spread to fixation in the absence of an effective male function. This may explain why a small but significant proportion of apomictic taxa are male sterile. For example, approximately 200 of the 1800 or so obligately asexual and non-pseudogamous *Taraxacum* agamospecies exhibit this condition (A. J. Richards, personal communication). The only reasonable explanation for this would seem to be that the ancestral apomict that founded each taxon was male sterile. There are two obvious ways in which an asexual individual may achieve greater reproductive success via the female function than a sympatric sexual individual. In the first, a switch to asexuality could be accompanied by a reallocation of resources from the male to the female function. In the second, the allocation of resources may not change but the ability of an asexual female to generate progeny via the female function may increase. Perhaps the first of these two mechanisms is the least likely. There is no obvious reason why a switch to asexuality should be accompanied by a change in allocation strategy. However, the second mechanism may be achieved whenever seed set in a sexual individual is limited by pollen availability (a frequent occurrence, Bierzychudek, 1981; Michaels and Bazzaz, 1986) rather than by the level of resources allocated to the female function. Removing the dependency on pollen will release an asexual mutant from this external constraint and will result in it maturing a higher proportion of its seeds than its sexual neighbours.

On the other hand, plausible situations can be envisaged in which the probability of an asexually produced individual surviving to reproductive maturity will, on average, be less than that of a sexually produced individual. For example, the progeny of asexual lineages may be less well adapted than those of sexual lineages to a changing environment. This will be discussed further in Chapter 6, but it is mentioned here to illustrate that the value of f could plausibly be less than, as well as greater than, 1. Under this circumstance, a gene for asexuality will only increase in frequency if there is an effective male function. But once fixation is achieved this function may become a costly and redundant feature, especially in non-pseudogamous obligate apomicts. At this time genes for male sterility would be selectively advantageous. But in the absence of gene flow between individuals, for a population to become male sterile, such a gene would have to arise independently in each apomictic individual, or would have to have arisen independently in a direct ancestor of each individual. This will be extremely unlikely. Thus the selective advantage of an effective male function during the establishment of asexuality, and the difficulty of losing it following establishment, may explain why many non-pseudogamous obligate apomicts produce pollen.

In conclusion, the male function can play an important, and sometimes a crucial, role in the spread of a gene for asexuality in cosexual heterosporous organisms. However, if the switch to asexuality is accompanied by an increase in the reproductive success of the female function (compared with sexual individuals), the role of the male function is reduced to one of enhancing an already existing advantage to asexuality. Under this circumstance, asexuality would spread to fixation, albeit more slowly, even if the asexual mutant was male sterile.

It is pertinent to move on now to examine the fate of a gene for asexuality that arises in homosporous plants. In order to do this, we must begin by questioning whether the assumptions made in the models presented above for the dioecious and cosexual heterosporous cases are valid for the homosporous case.

A major assumption of the models described so far is that a switch from sexual to asexual reproduction does not affect the number of eggs produced by an individual. It has been shown how this assumption is a valid one for animals and heterosporous seed plants. But it is not valid for homosporous plants. In order to see this, it is necessary to refer back to the description of the events associated with the Döpp-Manton and the Braithwaite schemes of agamospermy provided in Chapter 2 (Fig. 2.4 and accompanying text). In both cases, though by different pathways, each sporangium of an asexual individual produces only half the number of meiospores as a sporangium of a sexual individual. If a switch to asexuality simply affects the pathways by which meiospores are produced, an asexual mutant will produce only half the number of gametophytes, and consequently only half the number of progeny, as a sexual type.

The fate of an asexual mutation in homosporous organisms can be investigated by methods analogous to those used by Maynard Smith (1971, 1978) for dioecious organisms and by Charlesworth (1980) for cosexual heterosporous organisms. Throughout, it will be assumed that the gene for asexuality is dominant, and that asexual individuals are heterozygous for it and for the wild-type gene coding for sexuality. The fate of the gene for asexuality can then be followed simply by following the fate of asexual organisms, as all copies of the gene are contained in these, but each asexual organism possesses only a single copy.

In the first model to be presented, it will be assumed that the population comprises N sexual and n asexual individuals each of which produce the same number of sporangia (there is no reason to suppose that a switch to asexuality will affect the number of sporangia produced by an individual). As indicated in Chapter 2, each sporangium produces twice as many meiospores in a sexual individual than in an asexual individual (in many species the change is from 64 meiospores per sporangium in sexual individuals to 32 in asexual individuals, Fig. 2.5). If the probability that a meiospore produces a gametophyte is the same for sexual and asexual species then the number of gametophytes produced by a sexual individual can be represented by k and the number produced by an asexual individual by $0.5k$. Each gametophyte exhibits both a male and a female function. For reasons that will soon become apparent, it is best to deal with these separately. Let T represent the number of offspring successfully generated by each gametophyte via the female function. Then the proportion of the progeny generation produced by asexual individuals via this function is $0.5Tkn/(TkN + 0.5Tkn)$, or $0.5n/(N + 0.5n)$. If asexual individuals are unable to act successfully as male parents, this represents the proportion of the progeny generation that is asexual. Thus, under these assumptions, the proportion of asexual individuals in the population changes in one generation from

$$\frac{n}{N + n} \text{ to } \frac{0.5n}{N + 0.5n} \tag{3.6a}$$

When n is small relative to N, this approximates to a halving in the frequency of asexual individuals, or to a 50% cost of asexuality!

If asexual homospores are able to reproduce sexually via their male function, some or all of the cost of asexuality will be recouped. If it is assumed that all gametophytes are equally successful via their male function, the asexual component of the population will fertilize a proportion of eggs equivalent to the proportion of asexual gametophytes in the gametophyte pool. This differs from the cosexual heterosporous case described by Expressions (3.4), where the reproductive success of asexual individuals via their male function was equal to the proportion of asexual individuals, not to that of their gametophytes, in the population. This difference arises simply because the switch to asexuality does not affect gametophyte production in heterospores, but halves it in homospores. Thus in homospores, the asexual component of the population will sire a proportion $0.5kn/(kN + 0.5kn)$, or $n/(2N + n)$, of the TkN progeny generated by the sexual component. Each progeny sired by an asexual individual will be asexual. This arises because the male gametes are meiotically unreduced, being produced mitotically by the same meiotically unreduced gametophytes that exhibit the female function. Each male gamete will consequently carry a copy of the asexual mutation, and each progeny they contribute to will be asexual. This is another radical departure from the cosexual heterosporous case, where only half the male gametes carried a copy of the gene for asexuality, as these gametes were produced from reductional meioses. Thus, taking both the male and female functions into account, the proportion of asexual individuals changes in one generation from

$$\frac{n}{N + n} \text{ to } \frac{0.5n + N(n/\{2N + n\})}{N - N(n/\{2N + n\}) + 0.5n + N(n/\{2N + n\})}$$

or from

$$\frac{n}{N + n} \text{ to } \frac{0.5n + N(n/\{2N + n\})}{N + 0.5n} \tag{3.6b}$$

When n is small relative to N, $N(n/\{2N + n\})$ approximates to $0.5n$, and Expression (3.6b) approximates to

$$\frac{n}{N + n} \text{ to } \frac{n}{N + 0.5n} \tag{3.6c}$$

The cost of asexuality is overcome and is replaced by a very very slight advantage.

The substitution of the cost of asexuality with a slight advantage is due to the asexual component of the population achieving reproductive success through its male function and to the fact that the male gametes are meiotically unreduced, and hence each carries a copy of the mutation for asexuality. Unfortunately, although the male function fulfils the cost-cutting role superbly in the model, in reality it carries within it the seeds of failure. A cross between a sexual female and an asexual male will generate progeny that have a higher ploidy level than the asexual father, as each progeny receives a full complement of paternal chromosomes and a half complement of maternal chromosomes. Thus a cross between a diploid asexual paternal parent and a diploid sexual maternal parent will generate asexual triploid progeny, a cross between a triploid asexual paternal parent and a diploid sexual maternal parent will generate asexual tetraploid progeny, and so on. As a consequence, ploidy levels in lineages containing asexual male parents will increase until they reach levels that are deleteriously high, retarding the spread of a gene for asexuality via the male function. With the failure of the male function, the fate of the gene for asexuality will be totally dependent on the female function. As we have seen, there is a 50% cost to asexuality when only the female function in asexual individuals is effective. The consequences of the mode of action of the male function are illustrated in Table 3.2. It can be seen from this that the mean ploidy level of the asexual component of the population increases until the maximum ploidy level conducive with viability is attained. At this stage, the cost of asexuality is reimposed, and the asexual mutation begins to disappear from the population. Thus there is a kind of ratchet mechanism, due to the polyploidization effects of the male function, such that asexuality becomes concentrated at ever higher ploidy levels, becoming absent from lower ploidy levels.

Two factors which will affect the fate of a gene for asexuality in homosporous plants have been investigated in the model presented above – the 50% reduction in gametophyte production, and the deleterious consequences of increasing ploidy level following the fertilization of eggs of sexual individuals with male gametes from asexual individuals. However, the latter has not been examined in a rigorous quantitative way. Nor have we yet considered one other factor which may affect the fate of the gene – the consequences of it arising in a genotype of inferior fitness. These factors will now be considered (Mogie, 1990). The variable B will be used to provide a measure of the reduction in fitness due to the polyploidization effects of the male function $(0 < B < 1)$. As in Equations (3.1)–(3.3) and (3.5), individuals will be allocated fitnesses which are influenced by whether or not they are inbred. However, in addition, the fitness of an individual will also be influenced by whether or not it is asexual. Sexual individuals will be allocated fitnesses 1 and $(1 - d)$, with the reduction in fitness (d) being due to inbreeding depression $(0 \leqslant d \leqslant 1)$. The 50% reduction in gametophyte production that results from asexuality will make the corresponding fitnesses among the asexual component of the population 0.5 and $0.5(1 - d)$. As in Equations (3.5), the population will be assumed to comprise a proportion x of asexual individuals in one generation and a proportion x' in the next. This parameter will be broken down into parts x_1 and x_2 (where $x_1 + x_2 = x$) when the case

Table 3.2 This table depicts the drift to higher ploidy levels of the asexual component of a population of homosporous plants. The figures within the table describe the proportion of the asexual component at a particular ploidy level in a particular generation. The calculations assume that initially the population comprised a diploid sexual component and a diploid asexual component. Equal reproductive success via the asexual female and the sexual male function is assumed for the asexual component, as is essentially the case in Equations (3.6). Asexual reproduction produces progeny at the same ploidy level as the mother. Sexual reproduction involving a male gamete donated by an asexual sire, and a haploid egg produced by a sexual mother results in progeny that are one ploidy level higher than the sire, as the male gametes are meiotically unreduced

Generation	Ploidy level							
	2n	3n	4n	5n	6n	7n	8n	9n
1	1.0	0	0	0	0	0	0	0
2	0.5	0.5	0	0	0	0	0	0
3	0.25	0.5	0.25	0	0	0	0	0
4	0.125	0.375	0.375	0.125	0	0	0	0
5	0.0625	0.25	0.375	0.25	0.0625	0	0	0
6	0.03125	0.15625	0.3125	0.3125	0.15625	0.03125	0	0
7	0.015625	0.09375	0.234375	0.3125	0.234375	0.09375	0.015625	0
8	0.0078125	0.0546875	0.1640625	0.2734375	0.2734375	0.1640625	0.0546875	0.0078125

of the asexual mutant arising in an inbred individual is considered. x_1 represents that portion of asexual individuals that are inbred, and which consequently have fitness $0.5(1 - d)$. x_2 represents that portion of asexual individuals that are outbred, with fitness 0.5. It will be assumed that embryo and male gamete production per gametophyte is the same for sexual and asexual forms, and that the same proportion of male gametes is contributed by each gametophyte to the gamete pool involved in outcrossing. Each male gamete produced by the asexual component of the population is meiotically unreduced and a cross between an asexual paternal parent and a sexual maternal parent results in asexual progeny. Finally, it will be assumed that a fraction a of the matings in the sexual component of the population are between close relatives (leading to inbreeding depression) and that $(1 - a)$ are between unrelated individuals.

First, consider the case of a mutation for asexuality that occurs in a population that contains both inbred and outbred sexual and asexual members. The inbred asexual individuals will provide a proportion $0.5x_1 (1 - d)$ of inbred asexual individuals to the progeny generation via their female function. They will also produce asexual progeny via their male function, but as this involves mating with the sexual component of the population these progeny will be outbred. The sexual component of the population comprises a fraction $(1 - x)$ of the population. Of the eggs these produce, a proportion $(1 - a)$ is available for outbreeding. A proportion $0.5x$ of these will be fertilized by asexual sires but a fraction B of these will be inviable due to high ploidy levels. Thus the asexual component of the population will provide a proportion $0.5x(1 - B) (1 - x)(1 - a)$ of outbred asexual individuals to the progeny generation via the male function. Outbred asexual individuals will also be contributed to the progeny generation via the female function of outbred asexual parents. They will comprise a proportion $0.5x_2$ of the progeny generation. The fate of a gene will be determined by the mean fitness of its carriers compared with the mean fitness of the population as a whole.

The relative mean fitness of the population is 1 minus reductions in fitness due to the presence of the asexual gene and of inbreeding depression. The sexual component of the population produces a fraction $a(1 - x)$ of inbred progeny. The fitness reduction due to these is $ad(1 - x)$. $0.5x_1$ inbred asexual progeny are produced. These reduce fitness by $0.5dx_1$. The reduction in fitness due to asexuality is $0.5x$, and that due to polyploidization is $0.5Bx(1 - x)(1 - a)$. Thus the mean relative fitness of the population can be described as

$$1 - 0.5dx_1 - 0.5x - 0.5Bx(1 - x)(1 - a) - ad(1 - x)$$

or

$$1 - 0.5(x + x_1 d) - 0.5Bx(1 - x)(1 - a) - ad(1 - x)$$

Taking all of these factors into account, the following recurrence relations are given

$$x'_1 = \frac{0.5x_1(1 - d)}{1 - 0.5(x + x_1 d) - 0.5Bx(1 - x)(1 - a) - ad(1 - x)} \tag{3.7a}$$

and

$$x'_2 = \frac{0.5x_2 + 0.5x(1 - B)(1 - x)(1 - a)}{1 - 0.5(x + x_1 d) - 0.5Bx(1 - x)(1 - a) - ad(1 - x)} \tag{3.7b}$$

Combining Equations (3.7a) and (3.7b) gives x'. When x is so small that second order terms can be neglected, x' is approximated by

$$x' = \frac{x(1 - 0.5a - 0.5B\{1 - a\})}{1 - ad} \qquad (x \approx 0) \tag{3.7c}$$

The change in frequency of the asexual mutation is

$$\frac{x'}{x} = \frac{1 - 0.5a - 0.5B(1 - a)}{1 - ad} \qquad (x \approx 0) \tag{3.7d}$$

When there are no inbred asexual individuals in the population, $x_1 = 0$ and $x_2 = x$. The recurrence relation is taken from Equation (3.7b), making the necessary adjustments for the values of x_1 and x_2.

$$x' = \frac{0.5x + 0.5x(1 - B)(1 - x)(1 - a)}{1 - 0.5x - 0.5Bx(1 - x)(1 - a) - ad(1 - x)} \tag{3.7e}$$

When x is so small that second order terms can be neglected, Equation (3.7e) is the same as Equation (3.7c). Thus, the initial fate of the gene for asexuality is the same whether it originates in an inbred or an outbred individual.

Returning to Equation (3.7d), it can be seen that in the absence of deleterious ploidy effects (i.e. $B = 0$) but in the presence of inbreeding and inbreeding depression (i.e. $a > 0$, $d > 0$), a gene for asexuality will spread when $1 - 0.5a > 1 - ad$, which is when $d > 0.5$. That is, the gene will increase in frequency, only if the sexual component experiences very severe inbreeding depression. Deleterious effects due to polyploidization (i.e. $B > 0$) further limit the conditions under which the gene will persist in the population. However, in the absence of inbreeding (i.e. $a = 0$, $d = 0$) and of deleterious effects due to polyploidization (i.e. $B = 0$), Equation (3.7d) reduces to $(x'/x = 1)$. The gene for asexuality is neutral with respect to selection.

The right hand side of the numerator of Equation (3.7e) describes reproductive success by asexual individuals via their male function.

When this function is absent or ineffective then, when x is small and when the sexual component of the population experiences no reduction in fitness due to inbreeding, this equation can be approximated by $(x'/x) = 0.5$, demonstrating a 50% cost to asexuality.

The model described by Equations (3.7) demonstrates that a gene for asexuality will typically be selected against in a homosporous population. Only in the absence of inbreeding, of inbreeding depression and of deleterious effects due to polyploidization will the gene be selectively neutral. Within the confines of the model it never becomes selectively advantageous. Deleterious effects due to polyploidization will be felt, if not immediately, then after only a few generations (Table 3.2). Thus even if the gene is initially selectively neutral, and even if it initially increased in frequency through random genetic drift, selection pressures against it would soon appear and would quickly become very severe.

Of course, Equations (3.7) are derived on the assumption that the switch from sexual to asexual reproduction has no effect on spore viability, the number of sporangia produced, or the allocation of resources between the male and female functions. Altering these factors could reduce some of the costs of asexual reproduction. However, there is no indication from the literature that any of these ameliorating factors have been utilized by asexual homospores. Nor should they be expected. The gene for asexuality modifies the course of meiosis (Braithwaite scheme) or of the premeiotic mitosis (Döpp-Manton scheme), as described in Chapter 2. It is unlikely that it will also affect either sporangium initiation or the strategy of allocation between the male and female functions. Such changes would have to be achieved after the transition to asexuality, by the accumulation of further mutations. It is very unlikely that an asexual mutation would persist in a population for long enough for it to become associated with these extra mutations. It is possible that, because an asexual form produces 50% fewer spores per sporangium than a sexual form, asexual spores could be better provisioned than sexual spores, increasing their viability. There is no information on this, although there is evidence that the spores of some asexual forms may be larger than those of related sexual forms. This will be reviewed in the following section. Certainly, any increase in viability resulting from a size increase will have to be very large if the cost of asexuality is to be overcome.

An examination of asexual reproduction in cosexual homospores indicates that none have been able to overcome the cost of asexuality. The only obligately asexual homospores appear to be descended from female-sterile ancestors. Re-initiating a female function by switching to asexual, apogamous reproduction will have beneficial fitness effects in this circumstance.

3.3 THE CAPACITY FOR ASEXUAL REPRODUCTION IN HOMOSPORES AND HETEROSPORES

Section 3.2 provided some insights into why the distribution of asexuality among the bryophytes and tracheophytes is uneven: asexuality in homospores is only selectively advantageous when it provides an alternative to female-sterility, whereas in heterospores it is a selectively advantageous alternative to sexual reproduction. However, this knowledge clears up only some of the confusion noted in Table 3.1. For example, it tells us why, of the two groups in which asexual reproduction is common, facultative asexual reproduction should only be found among heterosporous angiosperms. This pattern of reproduction requires the retention of the capacity to reproduce sexually via the female function. A gene for asexuality would be selected against under these conditions in homosporous organisms. However, it does not tell us why the asexual production of embryos by apogamy has failed to become established in bryophytes, as female sterility must be a recurrent phenomenon in this taxon as it is among homosporous ferns where apogamy is widespread. Nor does it tell us why asexual reproduction has not been selected as an alternative to sexual reproduction in gymnosperms and heterosporous pteridophytes.

As well as failing to provide solutions to these problems, the models unearth a hitherto unrecognized but nevertheless potentially major barrier to the establishment of asexuality in homospores. A female-sterile member of an otherwise fully sexual homosporous population will be much less fit than its neighbours. Within the confines of the models described by Equations (3.6) and (3.7), it will have approximately half the fitness of these, this being achieved through its male function. Acquiring the capacity for asexual reproduction will restore its fitness to the level of its neighbours. However, this state will only be achieved temporarily, as asexual homospores produce meiotically unreduced male gametes. The deleterious consequences of this have already been explored: reproduction via the male function will result in an increase in ploidy levels in each generation, until these levels become deleteriously high, at which stage the male function becomes effectively redundant. A major implication of this has not been sufficiently stressed. This is that the acquisition of the capacity for asexual reproduction will not only provide only a fleeting reprieve to a female-sterile member of a cosexual, sexual population but will rapidly and effectively transfer the burden of reproduction from the sexual male to the asexual female function. The increase in fitness generated by the acquisition of asexuality will persist only during this transition period. Consequently, a gene for asexual reproduction will only persist in the long term if its carriers become isolated from sexual relatives.

This intuitively appears to be a requirement that is largely unattainable. And yet the existence of a large number of asexual fern taxa indicates otherwise.

There are two other problems which are not addressed by the models of the previous section. The first is concerned with why occasional asexual reproduction is practised by some homospores that possess a viable sexual female function. The second is concerned with why this type of asexual reproduction is typically apogamous rather than parthenogenetic, even though viable eggs are produced.

In order to find solutions to these problems it is necessary to consider factors other than the inherent costs and benefits of asexual reproduction. This is the purpose of this section. Some of the problems will be addressed briefly and directly, others, through necessity, by a longer and more lateral approach.

Asexuality, bryophytes, pteridophytes and the role of hybridization

The absence of obligate asexual reproduction in the bryophytes may simply reflect a low incidence of female sterility in this taxon. However, there are two other possible explanations. The first is that female sterility may not be uncommon, but that it is very unlikely that an individual which acquires it will also acquire the capacity to reproduce asexually. The second is that asexual reproduction may be acquired by female-sterile individuals, but it is not acquired within an ecological framework that is conducive to its establishment and spread. In investigating these possibilities it is pertinent to contrast aspects of bryophyte and pteridophyte ecology, as the conditions that have prevented the establishment of obligate asexual reproduction in bryophytes have clearly been absent in pteridophytes.

To begin with, it is reasonable to assume that the incidence of female sterility is lower among bryophytes than among pteridophytes as hybridization, which is commonplace in pteridophytes and is a common cause of female sterility, is a rare phenomenon in bryophytes. Indeed, it is unknown among liverworts, and occurs only rarely among other bryophyte groups (Smith, 1978). Thus the majority of instances of female sterility among bryophytes must arise as a consequence of mutation, rather than of hybridization, whereas instances of female sterility among pteridophytes will arise as a consequence of both phenomena, and will therefore be more common.

The extra opportunities that hybridization provides within the pteridophytes for the generation of female sterility may help to explain why obligate asexuality is much more common within this taxon than within bryophytes, but it does not explain why it is completely absent from bryophytes. One possibility is that hybridization plays a greater role in the evolution and establishment of asexuality in homospores than that of simply generating female-sterile forms. For example, an asexual phenotype may be more likely to be generated from the milieu of an unstable hybrid genome than from that of a more integrated, non-hybrid genome. This may be because asexuality, as well as female sterility, is induced by hybridization, because hybridization favours the phenotypic expression of a gene for asexuality that is possessed by, but has lain unexpressed in, one of the parental populations, or because hybridization may generate the ecological conditions that favour the spread of an asexual mutation. The first of these explanations is by far the most tenuous and I do not wish to defend it. I am aware of no firm evidence that hybridization *per se* can cause a switch from sexual to asexual reproduction. Indeed, the available evidence firmly points to the capacity for asexual reproduction being conferred by mutations that arise at specific loci. This issue will be investigated in detail in Chapter 5. The second and third explanations overlap to a certain extent, but it is useful to approach them separately.

If it is assumed, as seems justified, that hybridization-induced female sterility is, to all intents and purposes, absent from bryophytes, and that asexuality is genetically controlled, then one is led inexorably to the conclusion that the evolution of a potentially adaptive pattern of asexual reproduction in this group will initially require the acquisition of a minimum of two separate mutations – one for female sterility and one for asexual reproduction. Moreover, these mutations must be acquired in a way that results in them being phenotypically expressed simultaneously, as the expression of either one without the other would be highly deleterious. Expression of only the first would result in a female-sterile sexual phenotype that could achieve reproductive success only through its male function. Expression of only the second would result in a facultatively asexual phenotype that produced meiotically unreduced gametophytes from which embryos would be generated asexually by apogamy and sexually by the fertilization of unreduced eggs. The lineages resulting from utilizing these gametes in sexual reproduction would soon polyploid themselves out of existence. Only extremely rarely will these two individually deleterious mutations come together in the same genome. To have them coming together in a way that ensures that they are only phenotypically expressed when they co-occur will be even rarer.

Even if the establishment of asexual reproduction in female-sterile individuals has been achieved on occasion within the bryophytes, an asexual lineage will become established only if it becomes isolated from the sexual population in which it arises. Otherwise it will experience the cost of asexual reproduction (compared with sexual reproduction) that is associated with homospory. Yet there is no obvious reason why the acquisition of mutations for female sterility and for asexual reproduction should alter the basic ecology of asexual forms sufficiently to enable them to become isolated from their sexual compatriots. These mutations alter the reproductive biology, not the vegetative ecology, of their carriers. Given these constraints, it is not surprising that obligately asexual reproduction has not been recorded in the bryophytes!

It is pertinent at this point to consider how homosporous pteridophytes have been able to establish asexual forms when bryophytes have not. The arguments presented above are as relevant to pteridophytes as they are to bryophytes in that they indicate that successful asexual pteridophyte taxa are unlikely to have arisen by the accumulation of separate mutations for female-sterility and for asexual reproduction. Their evolution and establishment appears therefore to have been largely dependent on another factor. This brings the discussion back to the link between asexual reproduction and hybridization in the ferns. One obvious consequence of hybridization is that the number of mutations required for the evolution of obligate asexuality can be reduced to one. Hybridization can generate the female-sterile backdrop against which a mutation for asexual reproduction must play if it is to be acclaimed by the hypercritical forces of natural selection. However, this does not advance the investigation very far. It is clear that, for asexuality to become established, female sterility and the capacity for asexual reproduction must be expressed simultaneously, and in a situation in which the asexual mutant is not co-existing with sexual forms. Can hybridization achieve these objectives, in addition to that of inducing female-sterility? It can tentatively be suggested that it can.

First, consider its role in the generation of female-sterile forms. This may be a rare consequence of hybridization, possibly so rare that its overall rate is as low as that expected for mutation induced female-sterility. However, whereas the mutation rate will remain fairly constant, the opportunities to hybridize, and consequently the rate of hybridization-induced female sterility, will fluctuate widely from area to area and from time to time. For example, during periods of massive migration (e.g. driven by cooling and warming of the climate) previously allopatric taxa will become sympatric and will cross-fertilize, resulting in episodes characterized by very high rates of hybridization. During some of these many female-sterile forms will emerge. Next consider whether the generation of these hybrids will coincide with the expression of a mutation for asexual reproduction. There are two ways in which this could be achieved. The first requires that the mutation is already present in heterozygous members of the parent populations, in which it behaves as a recessive. It will consequently be passed on to some of the hybrids at the very moment at which the act of hybridization is generating female-sterile forms. That is, some hybrids will acquire female-sterility and the gene for asexual reproduction at their initiation. If the mutation continues to act as a recessive in these then asexuality will not arise. The sterile hybrids will simply become extinct. However, one consequence of hybridization can be the disruption of the processes that control gene expression. It can also induce dosage effects, especially if the parental types are at different ploidy levels. Thus a gene for asexuality that has persisted, possibly for many generations, as a recessive deleterious mutation in the gene pool of one of the parental types could find expression when transferred to a hybrid genome, with the result that some female-sterile hybrids, at the time of their initiation, will acquire and express the capacity for asexual reproduction. The second way in which newly arisen female-sterile hybrids could acquire a gene for asexual reproduction is by mutation. Assuming a typical mutation rate, the chances of this occurring will be very low, perhaps averaging between one in 100 000 and one in 1 000 000 per hybrid. The chances of a female-sterile hybrid acquiring a dominant form of an asexual mutation will be even lower. Thus the probability of a new female-sterile hybrid acquiring the capacity for asexual reproduction in this way is very small. But, given the long history of the pteridophytes and their capacity to hybridize, it is conceivable that this event has been realized on numerous occasions.

It is finally necessary to consider whether the origin of asexuality through hybridization-associated events will avoid the ecological problems associated with its origin simply through mutational events. What is required here is for hybridization to result in the ecological isolation of asexual forms from their sexual parents, so that the cost of asexuality is avoided. From our knowledge of the behaviour of hybrids it is not unreasonable to argue that hybridization can fulfil this requirement.

Hybridization among fern taxa could produce forms that are genetically diverse and that are sufficiently different from their parental types to exhibit a different range of ecological requirements or amplitudes. Thus newly arisen asexual hybrids could occupy new ecological niches, within the confines of which they will be protected from the costs of associating with their sexual parental types. There is some evidence that this eventuality has arisen at a reasonably high frequency. This is fortunate as I can think of no other explanation for the prevalence of asexuality among these organisms. One piece of evidence relates to the fact that a switch from sexual to asexual reproduction removes a constraint that greatly reduces the capacity of sexual homosporous ferns to occupy certain environments. The great majority of sexual ferns have an absolute requirement for free-water during reproduction. Male gametes reach the archegonia by swimming through this, and are attracted to the necks of the archegonia by chemical attractants introduced into it by the archegonia. Free-water is not a requirement for apogamous reproduction by asexual forms. These will consequently be able to establish themselves in drier sites than related sexual forms. The lifting of this constraint has long been recognized as a phenomenon that will benefit asexual ferns. Indeed, there have been several claims that the circumventing of a requirement for free-water during reproduction, which effectively results in asexual forms becoming adapted to dry habitats, could be the primary reason for the evolution of asexuality in this group (e.g. Wagner, 1974). I am not claiming this here, and Walker (1979) has rightly criticized this interpretation, pointing out that many genera which contain asexual forms are not representative of xeric sites. The argument I am offering is that asexual reproduction is selected as an escape from female-sterility but that a mutation for asexuality is only likely to increase in frequency if its carriers become ecologically isolated from their sexual relatives. One obvious route to achieve this is for asexual forms to occupy sites that lack free-water. Consequently, asexual forms may be over-represented in xeric sites in areas where hybridization-induced female sterility, and consequently asexuality, is common. But there is no reason to assume that they will be over-represented, or indeed even present, in xeric sites located elsewhere, where hybridization is uncommon. That is, asexuality may be more likely to become established in areas where xeric sites are readily available for colonization, but the availability of such sites will not in itself favour the evolution of asexuality.

A move into more xeric sites is only one of several ways in which asexual forms will be able to avoid the deleterious consequences of cohabiting with sexual forms. As stated above, a hybrid can acquire ecological attributes that differ from those of its parents, and which enable it to occupy sites these would find unfavourable. One consequence of this is that a hybrid may occupy a wider area than that occupied by its parental types. There are many examples of this in the Plant Kingdom, included among which are several fern complexes comprising sexual and asexual forms. For example, asexual forms are more widely distributed than related sexual forms in *Pellaea*, the *Polypodium plumula* complex and *Bommeria* (Tyron, 1968; Evans, 1969; Gastonby and Haufler, 1976), although the reverse situation is found in *Pteris* (Walker, 1979).

In conclusion, it appears that hybridization may have a crucial role to play in the establishment of asexual lineages in homosporous organisms. There is no reason to doubt that asexual forms can be generated in the absence of hybridization, but it is unlikely that they will be equipped to escape from the sexual population in which they arise. As a consequence, they will suffer the full cost of asexuality and will plummet rapidly to extinction. Hybridization provides a short-cut to asexuality, as it can induce the female sterility that would otherwise have to arise through mutation. But equally importantly, the genetic disruption that may accompany it may result in the expression of previously recessive genes for asexuality, and may result in the emergence of novel ecological attributes that will allow asexual lineages to become established in isolation from their sexual relatives. The general absence of hybridization from bryophytes, and its high frequency in ferns, would therefore appear to provide sufficient explanation for the absence of obligate asexual reproduction in the former, and its high frequency in the latter.

Apogamy, parthenogenesis, progenesis and the evolution of heterospory

Although obligate asexual reproduction is absent from bryophytes, and although among homosporous ferns it is restricted to female-sterile forms, a number of sexual taxa of both groups can reproduce asexually on occasion, most frequently as a laboratory phenomenon but also as a natural one. Overwhelmingly, these episodes of asexuality are apogamous rather than parthenogenetic. This situation can be contrasted with that in heterosporous plants where regular (obligate and facultative), occasional, and artificially induced asexual reproduction is usually parthenogenetic. This anomaly is of considerable interest in its own right, but in attempting to understand it the investigator is rewarded with much more than is initially promised. The bonus offered is an insight into the considerable role ontogeny plays in helping to determine the pattern and taxonomic distribution of asexual reproduction.

Occasional asexual reproduction in sexual homosporous ferns can be induced by supplying gametophytes with exogenous sugar and ethylene (White, 1979). It can also be induced by precluding sexual reproduction, for example by watering gametophytes from below to prevent a film of water from linking antheridia to archegonia, thus preventing spermatozoids from reaching eggs (Walker, 1979). In bryophytes, occasional asexual reproduction is most likely to be associated with the diploid gametophytes of tetraploid species rather than with the haploid gametophytes of diploid species. Apogamous embryos may arise from protonema, stems or leaves (Smith, 1978).

Occasional asexual reproduction is not associated in homospores with the fundamental cost that is associated with obligate asexual reproduction. This cost accrues because obligately asexual forms must avoid meiotic reduction. As a consequence, they produce 50% fewer gametophytes, and hence 50% fewer embryos, than sexual relatives. However, sexual forms that occasionally reproduce by apogamy do so from meiotically reduced gametophytes – the process is genetically and numerically equivalent to haploid parthenogenesis in animals and heterosporous plants. Thus occasional asexual reproduction is not associated with a reduction in the number of gametophytes or embryos produced. However, it is associated with other considerable costs. There is a basic cost of apogamy, as an embryo produced by this process is much more vulnerable than embryos produced sexually within archegonia. It lacks the protection afforded to these by the maternal gametophyte. An apogamous embryo, in contrast to a sexually produced one, grows exposed from its inception, there being no archegonial covering or calyptra (Lal, 1984). There is also a cost of haploidy, as this condition will allow the expression of all recessive, deleterious sporophytic genes, will lead to the loss of heterosis, and will lead to problems when, at maturity, the sporophyte attempts to initiate meiosis in its spore mother cells.

Although the phenomenon in homospores of occasional asexual reproduction poses no problems of interpretation, the fact that it is achieved almost exclusively through apogamy, rather than through parthenogenesis, is problematic. Why is it that asexuality is manifested through apogamy in these organisms but is usually manifested through parthenogenesis in heterospores? Homospores exhibiting occasional asexual reproduction do not lack a supply of eggs with which they could initiate parthenogenetic reproduction. And yet apogamy is still apparently preferred, even though it appears to provide an inferior mechanism of asexual reproduction – parthenogenesis would result in embryos being located within that part of the gametophyte that has evolved to afford them maximum support (i.e. the archegonium). Possibly, the general absence of parthenogenesis in homospores indicates a general inability to evolve it. There is some evidence that this is the case.

The plant life cycle involves two distinct stages: the sporophytic stage which produces meiospores from spore mother cells, and the gametophytic stage which produces gametes. The factors triggering these stages have long been of interest to biologists. A major goal has been to understand how the switch between sporophytic and gametophytic growth is achieved. As the gametophyte typically exhibits only half the total number of chromosomes exhibited by a sporophyte, one possibility is that the ploidy level primarily determines which stage is produced. However, this is not the case, as many taxa generate gametophytes and sporophytes that exhibit the same number of chromosomes (e.g. through apospory, apogamy or the avoidance of meiotic reduction during meiospore production). Evidence is accumulating that the fundamental cause of the transition is the regulation of gene expression through gross changes in the physiological and biochemical status of cells. Bell (1979a,b) provides an invigorating account, which will be summarized here, of these changes in homosporous and heterosporous ferns. (As described in Chapter 2, most ferns are homosporous, but heterospory is found in eight extant genera.)

In homosporous ferns, the transition from sporophytic to gametophytic growth is initiated at the spore mother cell stage through an episode of cytoplasmic autophagy (self-digestion of cytoplasm). The simpler metabolic environment this generates is thought to favour the expression of gametophytic genes and the suppression of sporophytic genes, enabling the meiospore, on germination, to give rise to the gametophytic stage of the life cycle. The transition from gametophyte to sporophyte is initiated in eggs. These are produced just behind the gametophyte's meristem. This area is metabolically rich as it lies in the flow of nutrients proceeding to, and of growth regulators proceeding from, the meristem. Due to its favourable location, the egg is able to develop an enriched cytoplasm, such that it contains more organelles, ribosomes and RNA than the meiospore. It also undergoes a period of enhanced genetic activity, evidenced by nucleocytoplasmic interactions during which tubular extensions of the nucleus push into the cytoplasm. The enhanced metabolic state of the egg is considered to confer 'femaleness' (Bell, 1979a) on it and to be conducive to the activation of sporophytic genes and the suppression of gametophytic genes – the egg is primed to generate a sporophyte.

The evolution of heterospory in the ferns involved the development of a radically new type of meiospore, the megaspore, which gives rise to the female gametophyte. Whereas the microspores of heterosporous ferns (which give rise to male gametophytes) are similar to the meiospores of homosporous ferns, the megaspores are much larger. For example, that of *Marsilea vestita* is some 2000 times the volume of its microspore. However, size is not the only characteristic that distinguishes a megaspore of a heterosporous fern from a meiospore of a homosporous fern. The megaspore is also much more metabolically active than the meiospore, and exhibits many of the characteristics of the homosporous egg. Thus the megaspore is endowed with a rich cytoplasm and undergoes the prolonged and extensive nucleocytoplasmic interaction that is associated with oogenesis in homosporous species. However, the eggs of heterospores generally lack the intense nucleocytoplasmic interactions found in the megaspore and in the homosporous egg. Overall, it seems that the development of the megaspore is in many ways analogous to that of the homosporous egg, rather than to that of the homosporous meiospore. Unlike the homosporous meiospore, but like the homosporous egg, the megaspore is primed for femaleness and sporophytic development and passes this potential onto its mitotic products, including its egg or eggs. As expected, only enough mitotic products are produced to allow this potential to be realized. Thus an archegonium, containing an egg that is already primed to generate a sporophyte, is initiated almost immediately after the megaspore germinates.

It is apt to enquire why the sporophytically primed megaspore undertakes to produce even a minimal gametophyte, instead of simply behaving directly as an egg. Presumably, this occurs because an embryo derived from a megaspore that adopted the role of an egg would be inviable, as it would lack the quasi-maternal support provided to embryos by the gametophyte. This is easily appreciated if heterosporous ferns are considered. In these, the megagametophyte is typically dispersed from the maternal sporophyte before or just after its egg or eggs have been fertilized. The immature embryo is incapable of supporting itself and its survival is completely dependent on it being associated with a gametophyte. Even in seed plants, where the female gametophyte (embryo sac) and its embryo are typically not dispersed from the maternal sporophyte until the embryo has reached, or is approaching, maturity, the female gametophyte undertakes a major role in embryo support, either directly, as is the case in gymnosperms, or by contributing two of its nuclei (the polar nuclei) to the endosperm initial, as is the case in most angiosperms. Thus any mutation that caused a megaspore to initiate an embryo rather than a gametophyte would be highly deleterious.

This comparison of homospory and heterospory in ferns would seem both to reinforce existing problems concerning the distribution of parthenogenesis and apogamy and introduce new ones. Although the timing of the transition from gametophytic to sporophytic potential varies between homospores and heterospores, the eggs of both groups are imbibed with sporophytic potential. However, the vegetative cells of megagametophytes also exhibit sporophytic potential, but those of the homosporous gametophyte do not. It would seem from this that both groups should be able to undertake parthenogenetic reproduction but that only heterospores should be able to undertake apogamous reproduction! Fact obviously contradicts supposition here.

To deal with the conundrum concerning parthenogenesis first, an important aspect of Bell's (1979a) investigations has been the demonstration that the almost complete absence of parthenogenesis from homospores is not due to their eggs lacking sporophytic potential. The explanation lies elsewhere, its location being within the egg nucleus.

Briefly, if the egg is to begin to divide mitotically it must first condense its chromatin sufficiently to meet the mechanical requirements of mitosis, as uncondensed chromatin will fail to align properly on the metaphase plate and will fail to pass, or will pass unequally, to the spindle poles. But the chromatin of the homosporous egg is very finely dispersed, so finely that it cannot condense unaided. It requires a template, the source of which is the much more condensed chromatin of the male gamete. With the help of the chromatin of the male gamete, the egg chromatin condenses to a stage compatible with a successful mitosis, and both sets of DNA become incorporated within the same nucleus. The resulting zygote enters mitosis to initiate the sporophyte. Thus there is an absolute requirement for fertilization in homosporous ferns, simply because a donor nucleus is required to allow the egg to realize its sporophytic potential by providing it with the means to undertake mitosis successfully! This explains the absence of parthenogenesis in this group.

The eggs of at least some heterosporous ferns are not subject to this constraint. For example, the chromatin of the egg of *Marsilea* is less dispersed than that of a homosporous egg and is capable of condensing without the aid of a spermatozoid template. Parthenogenesis is therefore possible and it has been observed in this taxon (Shaw, 1897). Insufficient is known of oogenesis in other heterosporous ferns to be able to predict with confidence whether the capacity for parthenogenesis observed in *Marsilea* is a general characteristic. Certainly, it is not restricted to *Marsilea*, as parthenogenesis has also been observed in *Selaginella* (Parihar, 1965). However, accounts of parthenogenesis are uncommon, and no heterosporous fern reproduces regularly by this process. Consequently, the conclusion to be made from Bell's investigations is that the evolution of heterospory in the ferns has been accompanied by the evolution of the capacity for parthenogenesis, but it is not yet clear whether these two phenomena are, like the Castor and Pollux of mythology, inseparable.

In the absence of a capacity for parthenogenesis, a homosporous fern must acquire alternatives if it is to reproduce asexually. Apogamy fulfils this role but, as we have seen, its presence is unexpected. An apogamously produced embryo must be initiated from a vegetative gametophytic cell that has acquired sporophytic potential, and yet in sexual taxa only eggs appear usually to acquire this. Bell (1979a) argues that this apparent inconsistency is due to differences between sexual and asexual forms in events associated with meiospore development. Meiospore development in asexual homosporous ferns is, to a significant extent, more closely analogous to megaspore development in heterosporous ferns than to meiospore development in sexual homosporous ferns. This predisposes asexual homosporous ferns to apogamy.

Thus he argues that the meiospores of asexual homosporous ferns should be larger than those of sexual relatives, as an asexual homospore produces only half the number of meiospores per sporangium as a sexual relative. He further argues that the cytological events associated with meiospore production in asexual homosporous ferns, including cytoplasmic autophagy, may be less severe than those observed in sexual forms, possibly because of the larger size attained by the meiospores of asexual forms. As a consequence of this, the capacity for sporophytic growth is not wholly eliminated. It will be realized in that part of the gametophyte which is most enriched by metabolites. This is the area immediately behind the apical meristem, where archegonia would be produced in sexual forms. In the absence of archegonia, vegetative cells in this area reach a metabolic condition in which the full expression of the sporophytic tendency is triggered, leading to the apogamous production of embryos. It must be stressed that this explanation of the control of apogamy is more speculative than that of the control of parthenogenesis, and its verification must await more detailed studies. However, there are a number of indications that it is fundamentally correct. For example, the meiospores of obligately apogamous *Dryopteris borreri* have a diameter that is between 20 and 25% greater than that of the related sexual species *D. filix-mas*. The extra volume that results from this is occupied principally by cytoplasm. There is also evidence that the cytoplasm of the spore mother cells of *D. borreri* undergoes less autophagy than that of the sexual species. Probably as a consequence of the extra cytoplasm acquired and retained by the meiospore, apogamous gametophytes mature faster, and generally produce embryos earlier, than those of related sexual forms (Whittier, 1970). Finally, supplying the gametophyte with excess levels of nutrients induces apogamy in sexual homosporous ferns (White, 1979). If vegetative cells of the gametophyte of an asexual form can acquire the nutrients which in a sexual form would be directed towards archegonia, as suggested by Bell, then apogamy should follow.

The events described provide an important insight into how the taxonomic distribution of different mechanisms or components of asexual reproduction has been influenced by ontogenetic constraints. The central role that these may play in shaping the distribution of asexuality, or of certain types of asexuality, is becoming more apparent. For example, evidence is accumulating that the absence of viable parthenogenetic reproduction in mammals is due to an ontogenetic constraint, rather than to ecological or other non-ontogenetic causes. It appears that the mammalian egg lacks the full complement of genetic information required to produce a viable, mature embryo. The missing information is provided by the male gamete, constraining mammalian reproduction to a pattern that is overtly sexual (Surani *et al.*, 1986; McGrath and Solter, 1986). In the light of this, it is pertinent to extend the present discussion to include the other heterosporous plant groups – the gymnosperms and angiosperms – to see if their capacity for parthenogenesis is linked to the ontogenetic changes that accompanied their evolution. Bell (1979a) notes a number of similarities between heterosporous ferns and seed plants that indicate that this may be the case.

The gymnosperm egg shares many characteristics of the heterosporous fern egg. It appears to maintain a low metabolic profile at the time of fertilization and, with the possible exception of cycads, there is no evidence of well-defined nucleocytoplasmic interactions during its maturation (Bryan and Evans, 1956; Foster and Gifford, 1959; Raghavan, 1976; Bell, 1979a). Similarly, the angiosperm megaspore, like the fern megaspore, appears to be well supplied with cytoplasmic components and to be metabolically very active. The angiosperm megaspore, because of unequal meiotic divisions, usually contains more than a quarter of the cytoplasm of the megaspore mother cell, and may absorb the products of the three degenerating megaspores of a meiotic tetrad (Willemse and de Boer-de Jeu, 1981; Bouman, 1984; Willemse and Van Went, 1984). Based on the few quantitative data available (e.g. Russell, 1979), an increase in organelles, ribosomes and other cellular components occurs during the development of the megaspore, but not during oogenesis (Willemse and Van Went, 1984). Nucleocytoplasmic interactions of the type found in the megaspore of the heterosporous fern *Marsilea* may also occur in the megaspores of the angiosperm *Myosurus* (Woodcock and Bell, 1968a,b; Bell, 1979a). The egg appears to be relatively inactive, although only a few studies have been made (Willemse and Van Went, 1984).

However, although the few available data point to similarities between the heterosporous condition in ferns, gymnosperms and angiosperms, many more data need to be obtained from the latter two groups before any firm conclusions can be reached. One problem in particular needs to be resolved. An unstated assumption of equating events associated with heterospory in ferns with those associated with this event in seed plants is that the functionally equivalent components of the different heterosporous systems are homologous. Another is that the capacities and potentials of homologous structures are maintained through phylogeny. These assumptions can be questioned. For example, although it is reasonably clear that the megaspores of gymnosperms, angiosperms and heterosporous ferns are homologous, it is by no means certain that they exhibit the same potential. The female gametophytes of most gymnosperms, though much smaller than the gametophytes of homosporous organisms, are much larger

and have a much longer prereproductive stage than the female gametophytes of flowering plants and of most heterosporous ferns. Is this
because the gymnosperm megaspore, like the homosporous fern meiospore, lacks sporophytic potential, this state being conferred only after
a lengthy period of gametophytic growth? Or is this potential present
in the gymnosperm megaspore but, contrary to the situation in other
heterospores, suppressed? There is insufficient information available
to allow this issue to be resolved. However, its existence shows that
great caution must be exercised if comparisons are to be made between
the various groups of heterosporous organisms.

The doubts concerning the status of megaspores extend much more
forcibly to the status of the female gamete. The eggs of homosporous
and heterosporous ferns appear to be homologous, although they exhibit different potentials for parthenogenesis. However, a comparison of
archegonium development in ferns, gymnosperms and angiosperms
provides evidence that the eggs in these groups are not homologous,
the gymnosperm egg being homologous to a precursor of the fern egg,
and the angiosperm egg being homologous to another, non-gametic,
cell in the fern archegonium. This has implications for the study of
parthenogenesis, as replacing an egg cell that may not be capable of
dividing unless it is fertilized, with another cell from the archegonium
that does have this capacity, is an efficient, effective and parsimonious
means of introducing the capacity for parthenogenesis to a lineage. (It
is not being argued that replacement has been selected to enable
organisms to reproduce parthenogenetically. Rather, it is being suggested that parthenogenesis is a fortuitous consequence of replacement, which has evolved in response to other selective pressures. This
will be made clear below.)

The gymnosperm egg is homologous to precursors of the fern egg.
In some taxa the precursor is the central cell itself, in others it is the
primary ventral cell or its nucleus. This is shown in Fig. 3.1. The
homology between the gymnosperm egg and precursors of the fern egg
is of considerable interest as these precursors in ferns express mitotic
potential (Fig. 3.1) which may be expressed as parthenogenetic potential in the gymnosperms. It is more difficult to identify the fern cell
that is homologous with the angiosperm egg. It has been variously
identified as being the fern egg itself, an archegonium initial, or a neck
canal cell (see review by Favre-Duchartre, 1984, and Battaglia, 1988,
1989). However, it is unlikely that any single fern cell can be identified
as the sole homologue of the angiosperm egg as it is unlikely that
homology is maintained even within the angiosperms. Certainly, the
number of mitotic divisions in the megagametophyte that leads up to
egg differentiation varies between taxa, the minimum number being
one (e.g. the *Adoxa* type of eight-nucleate tetrasporic embryo sac)

and the maximum three (e.g. the *Polygonum* type of eight-nucleate monosporic embryo sac). A detailed account of variation in embryo sac formation is provided by Willemse and Van Went (1984). This variation does not pose too many problems, as one pattern of embryo sac formation is predominant within the angiosperms, and almost all apomicts exhibit it. This pattern is described as the eight-nucleate 'Normal' or *Polygonum* type. The small number of apomicts that do not exhibit it include *Ixeris dentata* (Okabe, 1932), *Rudbeckia* (Battaglia, 1946; Fagerlind, 1946), *Erigeron* (Fagerlind, 1947; Battaglia, 1950), *Statice oleaefolia* (D'Amato, 1949) and *Allium* (Hakansson, 1951; Hakansson and Levan, 1957). Attention will consequently be focused on this type. The most parsimonious interpretation of its ontogeny will be presented. It is based on Chadefaud (1941; reported in Favre-Duchartre, 1984) and is illustrated in Fig. 3.2. The embryo sac (megagametophyte) is considered to contain no vegetative cells (i.e. no cells that cannot be equated with a cell of the axial row of an archegonium). Through a process of progenesis, the vegetative megagametophyte is unicellular and comprises the megaspore itself. Its nucleus divides to give two nuclei, each of which is homologous to the central cell of an archegonium. Each of these divides twice to give the two sets of four nuclei which comprise the eight-nucleate embryo sac. Thus the mature embryo sac comprises the axial rows of two archegonia. These are inverted with respect to one another, so that their component nuclei take on the roles depicted in Fig. 3.2. The fern egg is homologous with the angiosperm polar nucleus. The angiosperm egg is homologous with the fern ventral canal cell. In this scheme, only one of the axial rows gives rise to an egg apparatus (comprising the two synergids and the functional egg). The equivalent nuclei in the other axial row give rise to three antipodal cells. This difference in behaviour can be explained as a phenotypic response to local conditions. Physiologically important gradients of various sorts are found within the embryo sac. That these can have a determining effect on the phenotypes of the embryo sac nuclei is illustrated by Hakansson's (1951) investigations into embryo formation in *Allium*. Very occasionally the three nuclei that usually differentiated as antipodals differentiated instead as an egg apparatus, the egg of which divided to form an embryo. Unfortunately, Hakansson's observations cannot be used to corroborate the scheme described, as embryo sac formation in *Allium* does not follow the *Polygonum* pattern. However, they do illustrate that cells within the embryo sac that have followed different patterns of differentiation may nevertheless be homologues.

This search for homologues is important. Within the axial row of the homosporous fern archegonium, only the egg cell is known to be typically unable to divide. As a consequence, the ability of seed plants to undergo parthenogenetic reproduction may be due primarily to them having adopted, as a functional egg, a cell that is non-homologous to the fern egg, and which has retained the capacity for mitosis. Interestingly, cells in the axial row other than the egg cell are known to be able to exhibit gametic tendencies in a wide variety of taxa. Adopting one of these cells as the functional female gamete may consequently involve no more than harnessing this tendency. Thus the ventral canal cell has been observed to fuse with the egg cell in

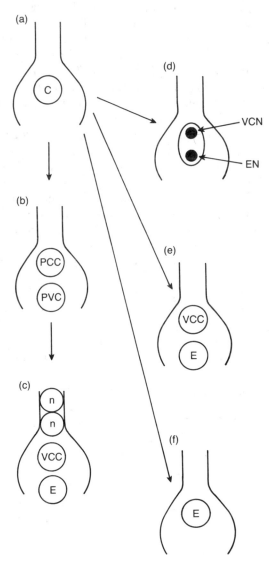

some bryophytes (e.g. *Sphagnum* species (Bryan, 1920)). This phenomenon has also been observed in a number of gymnosperms, including *Encephalartos villosus* (Sedgewick, 1924), cycads (Chamberlain, 1935), *Picea laricio*, *Abies balsamea* and *A. pindrow* (Dogra, 1966), although the gymnosperm ventral canal cell is homologous to the fern primary canal cell, not to the fern ventral canal cell. Embryos have also been observed to arise from synergids (which are homologous to fern neck canal cells) in flowering plants (Kimber and Riley, 1963). In addition, fertilization of the ventral canal cell has been observed in the gymnosperm *Ephedra* (Khan, 1943). The resulting 'zygote' may divide a few times before degenerating.

The cause of these substitutions in the seed plants almost certainly differs between gymnosperms and angiosperms. Pollination in the gymnosperms occurs long before (often several months before) fertilization, while the megagametophyte is still very immature. Indeed, in some species pollination may occur before the megaspore has begun to divide. As a consequence of this, the developing megagametophyte is surrounded by several male gametophytes which will compete to fertilize eggs as soon as archegonia mature. However, although the megagametophyte produces several archegonia, and thus several embryos, only one embryo will be allowed to mature. A male that adopts a precocious fertilization strategy may be more likely to be the sire of this embryo than a male that is more laggardly. Perhaps the substitution of the central cell or the primary ventral cell for the egg is a consequence of this, a successful male strategy being to fertilize a precursor of the egg rather than wait for this precursor to divide to produce the egg.

The situation in the angiosperms is very different. Pollination is delayed until after the megagametophyte has matured, precluding the use of precocious fertilization as a male strategy. As a result, the ontogeny of the axial row of the archegonium has not been abbreviated. Substitution has resulted because, uniquely among plants, the

---

Figure 3.1 The development of the axial row of an archegonium in ferns and gymnosperms. The pathway in ferns is depicted in (a)–(c). the central cell (C) in (a) divides to produce the primary canal cell (PCC) and the primary ventral cell (PVC) in (b). The primary canal cell initiates one or more divisions to give two or more neck canal cells (n) and the primary ventral cell divides to produce the ventral canal cell (VCC) and the egg (E) in (c). In gymnosperms, the central cell either differentiates immediately as the egg (steps (a) and (f)) or it undergoes a single mitosis to produce either a binucleate structure, the lower nucleus of which functions as the egg nucleus (steps (a) and (d)), or two daughter cells, the lower of which functions as the egg (steps (a) and (e)). Thus the egg in some gymnosperms is homologous to the fern central cell, whereas in others it is homologous to the fern primary ventral cell or its nucleus. EN = egg nucleus, VCN = ventral canal nucleus.

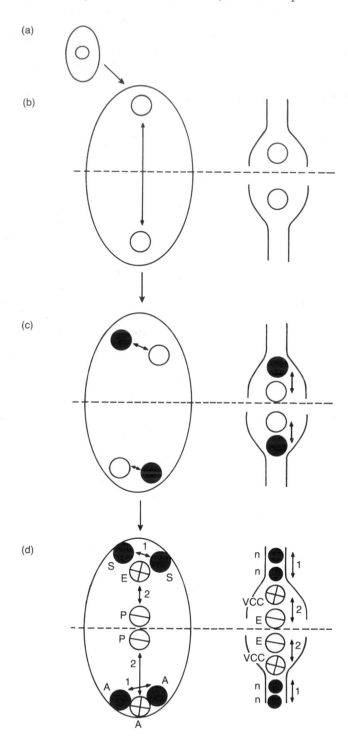

nurture of the embryo has been transferred from the maternal gametophyte to a new tissue, the endosperm, the initiation of which (in most
taxa) involves the fusion of two megagametophyte cells (the polar
nuclei) and a male gamete. In order to evolve endosperm the angiosperms have had to allocate cells with gametic potential to the role
of polar nuclei. The most parsimonious way to achieve this is to
allocate this role to the cell that is homologous to the fern egg (requiring the production of two archegonial axes if two polar nuclei are
required) and to confer the role of female gamete onto this cell's sister
cell – the homologue of the fern ventral canal cell. In so doing, the
endosperm is evolved without having to alter the framework of the
ancestral archegonial axis (shared by both ferns and seed plants as
these share a common ancestor in the Trimerophytopsida, Fig. 2.2)
and is automatically located in that part of the megagametophyte in
which it can operate most effectively.

Thus the substitution of the ventral canal cell for the egg in angiosperms is most likely to be a consequence of the switch in the pattern
of maternal provisioning of embryos. The substitution of the central
cell or the primary ventral cell for the egg in gymnosperms is most
likely to be the result of male-male competition.

This discussion of homology would be interesting, but not particularly relevant to obtaining an understanding of the distribution of
parthenogenesis if, as a consequence of the evolution of heterospory,
the capacity for parthenogenesis was acquired by the egg. We have

---

Figure 3.2   The proposed relationship between the development of the *Polygonum* type of embryo sac of flowering plants (left-hand side) and the axial
row of the homosporous archegonium (right-hand side). The meiospore which
in flowering plants is described as the megaspore, divides mitotically three
times to give the mature eight-nucleate embryo sac (d). In homospores, the
meiospore divides mitotically a number of times to produce the multicellular
gametophyte stage, cells of which eventually differentiate as central cells of
archegonia. (b) The angiosperm megaspore nucleus is shown dividing to give
two central cell nuclei, which are homologous to the central cells in each of
the two archegonia depicted on the right. The archegonia are inverted with
respect to each other. (c) The central cell divides to give two daughter products
which, in archegonia, are described as the primary canal cell (solid circle) and
the primary ventral cell (open circle). (d) The archegonial primary canal cell
divides to give neck canal cells (n, solid circles); the archegonia primary
ventral cell divides to give a ventral canal cell (VCC, open circle with cross)
and an egg (E, open circle with single bisector). The embryo sac homologues
of these are shown with the same markings. The archegonial egg is homologous to the embryo sac polar nucleus (P). The archegonial ventral canal cell
is homologous to the embryo sac egg (E, top part of embryo sac in (d)) or to
an antipodal cell (A, bottom part of embryo sac). The archegonial neck canal
cells are homologous to synergids (S, top part of embryo sac in (d)) or to
antipodals (bottom half).

seen that this capacity was acquired with heterospory in at least some fern lineages. However, the seed plants are not descended from these. But if it was also acquired with heterospory in the direct ancestors of the seed plants the changes described in the identity of female gametes in gymnosperms and angiosperms may lack great consequence. Even if another cell had not been substituted for the egg, these taxa would still express a capacity for parthenogenesis. However, although conclusive proof is unavailable, there is reason to believe that the capacity for parthenogenesis in seed plants has been greatly increased by substitution.

The importance of substitution cannot be determined in the gymnosperms, as these do not produce a cell that is homologous to the fern egg because of the abbreviated pattern of development of the archegonial axis. However, this matter can be investigated in angiosperms by comparing the capacity of the female gamete and the polar nuclei to divide in the absence of male gametes. In doing so, it soon becomes obvious that they have very different capacities. The female gametes of many taxa can divide parthenogenetically, either naturally or following artificial stimulation (Kimber and Riley, 1963), but in only a subset of these – autonomous apomicts – do the polar nuclei show this capacity, and only then usually after they have fused together. In sexual species, and in pseudogamous apomicts, a male gamete must fuse with the polar nucleus/nuclei (some sexual species and some pseudogamously apomictic species produce only a single polar nucleus in each embryo sac (Willemse and Van Went, 1984; Nogler, 1984) before endosperm is produced. Indeed, it is debatable whether the polar nuclei in those pseudogamous apomicts that produce two such nuclei can complete fusion in the absence of a male gamete. Certainly, this appears to be the case in a number of sexual species (Willemse and Van Went, 1984), where fusion between the polar nuclei, though initiated before fertilization, is not completed until the male gamete infiltrates the proceedings to give, at gametic level, a *ménage à trois*. Perhaps pseudogamous apomicts are descended from this category of sexual ancestor. Whatever the cause of the requirement for fertilization of the polar nuclei of pseudogamous apomicts, its presence forces the conclusion that, but for substitution, the incidence of apomixis in the angiosperms would be lower than it is. Parthenogenetic apomixis would be restricted to groups that have demonstrated they can evolve autonomous apomixis. As most of these belong to the Asteraceae (Nogler, 1984), which contains less than a third of the genera in which parthenogenetic apomixis has been recorded (Table 2.1), the taxonomic distribution of parthenogenetic apomixis would be very restricted indeed.

It is worth returning briefly to the pteridophytes, as a doubt was raised previously about whether the capacity for parthenogenesis was automatically or only occasionally acquired when heterospory was acquired in this group. The phenomenon of pseudogamous apomixis in the angiosperms indicates that only some heterosporous ferns may have acquired it. However, even if this is not the case, there is some doubt that the events associated with the acquisition of heterospory by the ferns have also been associated with its acquisition by other groups – at least to the same extent. Thus a major characteristic of the evolution of heterospory in ferns is the transfer of events from oogenesis in homospores to megasporogenesis in heterospores (Bell, 1979a). One of the most obvious manifestations of this is the transfer of nucleocytoplasmic interactions from the homosporous egg to the heterosporous megaspore. However, Wilms (1981) has shown that these interactions are characteristic of the polar nuclei of *Spinacia* (spinach). That is, the polar nucleus of *Spinacia*, which is homologous to the fern egg, behaves much more like the egg of a homosporous fern than the egg of a heterosporous fern. It is not known whether the *Spinacia* polar nucleus is typical of the angiosperms, but its existence demonstrates that great caution should be exercised when events described in one major taxon arc used to understand parallel events in another.

Precocity, pre-adaptation and asexuality in gymnosperms and angiosperms

One of the most persistent problems concerning the distribution of asexual reproduction in plants is the absence of gymnosperms that produce embryos only by asexual processes. Their absence is unexpected on several counts. Gymnosperms have been observed to reproduce occasionally by haploid parthenogenesis (Chapter 2) demonstrating that their eggs can divide without the stimulus of fertilization. They are heterosporous, and thus will experience the cost of sex if an asexual mutant arises. And there is nothing about their ecology to suggest that, if such a mutant arises, the fundamental advantage of asexuality will be overturned by environmental factors. And yet, regular asexual reproduction (e.g. facultative or obligate apomixis) has never been observed. This contrasts markedly with the situation in angiosperms, where regular asexual reproduction is a characteristic of many taxa. It could be argued that, as less than 1% of angiosperm species are asexual, the absence of asexual gymnosperms is to be expected as this class comprises only some 730 species. But this falls far short of the requirements of a valid explanation. I mention it only because I have heard it used as the latter. The argument is credible only if it can be reasonably assumed that the ability to evolve and maintain patterns of asexual reproduction is constant across the two classes. And yet it is not constant even within the angiosperms, as is shown in Chapter 2. Rather than simply and unthinkingly dismissing the absence of asexual taxa from the gymnosperms as a statistical artefact, a much more reasonable approach is to ask whether their absence may be due to some aspect of gymnosperm biology. In taking this approach much can be learned about the role of ontogeny in determining the distribution of asexuality.

Four conditions must be met before a transition from sexual reproduction to a regular system of parthenogenetic apomixis can be achieved in cosexual heterosporous seed plants. First, the capacity for parthenogenesis must be present; second, this capacity must be allowed expression by preventing male gametes from fertilizing the eggs; third, the eggs must exhibit the same ploidy level as the mother, so meiotic reduction during egg production must be avoided; fourth, the first three conditions must be met simultaneously, or almost so. This is because each is deleterious if expressed in the absence of the others. For example, if only the capacity for avoiding meiotic reduction is acquired by a lineage, the ploidy level will increase each generation as the unreduced eggs are fertilized, and will rapidly reach a level that is incompatible with reproductive success or with viability. Similarly, if all but the capacity for the avoidance of meiotic reduction is acquired, haploidy, and possibly subhaploidy, will result and the advantages of diploidy will be lost to the lineage. A major problem in meeting the fourth requirement is that if more than a single mutation is required for the first three conditions to be met, these must not only be acquired but they must be acquired in such a way that they become phenotypically expressed simultaneously, or at the very most within a very few generations of each other. Otherwise the problems described above associated with changes in ploidy level will be experienced.

The absence of asexual reproduction in gymnosperms cannot be due to them lacking the capacity to meet any of the first three conditions. Parthenogenesis is known to occur, and meiotic irregularities of the type associated with the avoidance of meiotic reduction in other organisms have been observed in gymnosperms (e.g. asyndesis – Andersson, 1947; Runquist, 1968; Baker *et al.*, 1976). The avoidance of fertilization has not, to my knowledge, been observed, but it is inconceivable that the female reproductive apparatus of gymnosperms is immune to disruption – as it is genetically determined, its capacity to capture male gametes and direct them towards eggs must be susceptible to undermining by mutation. It follows that the difference in the ability of gymnosperms and angiosperms to undertake the transition to regular parthenogenetic reproduction must be due to them exhibiting different capacities to fulfil the fourth condition – that of meeting the first three more or less simultaneously.

The probability of a lineage being able to fulfil the fourth condition will be negatively correlated with the number of mutations needed to fulfil the first three. It is therefore important to determine how many mutations may be required for the transition to asexuality in angio sperms and gymnosperms. The expectation is that fewer will be required by angiosperms. It is unlikely that either group will be required to generate a mutation to acquire the capacity for partheno-genesis. It has already been argued that this capacity is present in seed plants as a consequence of the changes in the function of the components of the axial row of the archegonium that accompanied their evolution. This assumption will be investigated at length in Chapter 5, where it will be justified. Consequently, it will be assumed that this capacity will be expressed by many taxa of both gymno-sperms and angiosperms if fertilization is avoided. However, it is likely that a mutation will be required by both groups if the reductional female meiosis is to be replaced by a mitosis or by a restitutional meiosis. Such changes have been intensively investigated in sexual organisms (Baker *et al.*, 1976) where it is clear that they are typically mediated by mutation. It is therefore unreasonable to assume a non-mutational origin in apomicts. This matter will also be investigated in much greater detail in Chapter 5 where, once again, the assumption being made here will be justified. With the capacities for parthenogen-esis and the avoidance of meiotic reduction being met in the same way by both groups, it appears that the difference in the capacities of angiosperms and gymnosperms to acquire regular asexual repro-duction must be due to them exhibiting different requirements for the avoidance of fertilization. This seems to be the case.

The simplest way to ensure that fertilization is avoided is physically to prevent male gametes from having access to the egg. One way to achieve this is to prevent mating. Another way is to extend the period between egg production and mating so that the egg has time to initiate an embryo parthenogenetically, or at least to develop to a stage at which it is no longer susceptible to fertilization, before male gametes are given access to the megagametophyte. Almost all apomictic flowering plants follow this second strategy. The way they achieve it is by maturing eggs precociously rather than by delaying pollination. The few taxa that do not follow this strategy exhibit hemigamy. Here, a male gamete nucleus enters the egg but fails to fuse with the egg nucleus and eventually degenerates. Fertilization is avoided because the male nucleus either fails to divide, or does so asynchronously with the female nucleus before degenerating after a few mitotic cycles.

Apomictic flowering plants that exhibit precocious egg maturation can be divided into two groups. In one, precocity is so advanced that the egg has already initiated an embryo before pollination. This phenomenon is described as precocious embryony. The second group comprises apomicts in which the egg has not divided by the time male gametes enter the embryo sac but has reached a stage in its development at which fertilization is apparently impossible. This phenomenon will be described as precocious oogenesis.

In apomicts exhibiting precocious embryony, the egg divides parthenogenetically before the flower has opened and, in cosexual forms, before the flower's anthers have dehisced and/or its stigmas have matured. Consequently, access to the egg is denied to both self pollen and foreign pollen. Precocious embryony is characteristic of many facultatively and pseudogamously apomictic taxa as well as of many obligately and autonomously apomictic taxa. The facultative apomicts initiate embryos parthenogenetically from meiotically unreduced eggs before pollination is possible and before meiotically reduced eggs have matured; the meiotically reduced eggs mature at a time when they can be fertilized, that is, when the flower is open and the stigmas are receptive to pollen. The pseudogamous apomicts initiate embryos parthenogenetically before pollination so that the embryo is already sizeable when male gametes finally gain access to the embryo sac; the polar nuclei are available for fertilization, but the egg no longer is.

Precocious oogenesis appears to be a rudimentary form of precocious embryony. Male gametes enter the embryo sac before the egg has divided but fail to fuse with it. The reasons for this failure are not clearly understood. The age or stage of development of the egg relative to that of the pollen may be important in preventing fertilization (Mogie, 1988). Unfertilized eggs have incomplete cell walls. In sexual taxa, the male gamete gains entry to the egg via the gap in the wall, after which the wall extends to close the gap. Wall formation may be a function of the age or stage of development of the egg, so that a precociously maturing egg may be completely enclosed by its wall before a male gamete can reach it, resulting in fertilization being avoided. Thus in these taxa the egg progresses to the stage where it is no longer accessible to male gametes, even though it has not progressed to the stage at which it will divide parthenogenetically.

Precocity provides a relatively simple mechanism for avoiding fertilization. It allows apomixis to become established without having to resort to the gross changes in behaviour or anatomy that would have to be acquired if mating (pollination) was to be avoided. Indeed, the ability of flowering plants to avoid fertilization through precocity, rather than through the avoidance of mating, has had an important enhancing effect on the incidence of apomixis in this group, as the evolution of pseudogamous apomixis is dependent on pollination being maintained. However, precocity will only provide a strategy for avoiding fertilization if pollination occurs after eggs have been produced. This is the case in angiosperms, but it is not the case in gymnosperms, where pollination usually occurs weeks or months before eggs are initiated. As a result of this, male gametes of gymnosperms are in a position to fertilize the eggs as soon as they are mature, preventing them from realizing any potential for parthenogenesis. The only strategy for avoiding fertilization available to gymnosperms is the prevention of pollination.

It is clear that gymnosperms and angiosperms face different problems in avoiding fertilization. Gymnosperms need to prevent pollination. Angiosperms need only effectively to delay it. This difference provides the key to understanding why regular parthenogenetic reproduction is common in angiosperms but absent from gymnosperms. Although angiosperms avoid fertilization by effectively delaying pollination, non-pseudogamous apomicts could, in theory, avoid it by avoiding pollination. To my knowledge, none have done so. This strongly suggests that the latter strategy is much more difficult to evolve than the former. Indeed, the absence of regular asexual reproduction in gymnosperms indicates that the avoidance of pollination must be almost impossible to acquire as a component of an asexual system. This is not to say that the avoidance of pollination cannot be achieved. It undoubtedly can be, although achieving it will involve considerable modifications to the structure of the pollen collecting surfaces (the stigma in flowering plants or the ovule and pollination drop in gymnosperms). Such modifications will require the accumulation of several mutations. It is this that makes this strategy an unsuitable one with respect to the evolution of asexuality. These mutations will have to be acquired and phenotypically expressed at about the same time as the capacity to avoid meiotic reduction is acquired and expressed. This complex juxtapositioning of these separate events is unlikely to be achieved rapidly enough. Certainly, it has not been achieved by any extant lineage in either the gymnosperms or the angiosperms.

So far, this comparison between gymnosperms and angiosperms, and within angiosperms, has shown that precocity is much easier to evolve than the avoidance of pollination, so much so that only those taxa whose reproductive behaviour provides them with the opportunity to opt for precocity as a means of avoiding fertilization have been able to evolve apomixis. The comparison has not told us why this is. In order to see this, it is necessary to look in detail at the events leading up to egg production in angiosperms. In doing this, it soon becomes obvious that precocity is, almost certainly, simply a consequence of the avoidance of meiotic reduction, rather than a phenomenon that needs to be acquired separately. The implication of this is that if parthenogenesis is an innate capacity, as seems to be the case, then a viable system of apomixis can evolve in angiosperms simply as a result of the acquisition of the capacity to avoid meiotic reduction. However, there is slightly more to the problem than this, as the way in which meiotic reduction is avoided will determine whether a viable pattern of asexual reproduction is generated. These issues will be illustrated by investigating the relationship between precocity and the avoidance of reduction in both generative and aposporous apomicts.

As discussed in Chapter 2, most generative apomicts, produce meiotically unreduced embryo sacs by the *Antennaria* scheme, although the *Taraxacum* scheme is common within the Asteraceae, and the *Ixeris* scheme is exhibited by a few taxa. The differences between these schemes are outlined in Fig. 3.3, which also provides an outline of the ancestral sexual schemes from which they evolved. The sexual *Polygonum* scheme is ancestral to the *Taraxacum* and *Antennaria* schemes, while two sexual tetrasporic schemes (the *Drusa* and *Fritillaria* types, Nogler, 1984) are ancestral to the *Ixeris* scheme. It can be seen from Fig. 3.3 that fewer stages are involved in the production of a mature embryo sac from a generative cell in apomicts than in their sexual ancestors. The sexual *Polygonum* scheme involves the embryo sac mother cell (the megaspore mother cell) undergoing the two divisions of a reductional meiosis to produce the megaspore, which initiates three mitotic cycles to form the mature eight-nucleate embryo sac. Thus a total of five cell divisions separates the generative cell from the mature egg. In contrast, only three divisions separate the generative cell from the mature egg in apomicts undergoing the *Antennaria* scheme. Meiosis is completely omitted in these. Instead, the generative cell functions directly as the megaspore, undergoing the three mitotic cycles to produce the embryo sac. The number of steps taken to produce the egg in these apomicts is reduced 'by a whole meiosis' over the number taken by their sexual ancestors. Fewer steps are saved in the *Taraxacum* scheme. Here, the generative cell undergoes both divisions of meiosis to produce a megaspore which undergoes three mitotic cycles. Thus there is no difference in the number of divisions separating the generative cell from the egg in the *Taraxacum* and *Polygonum* schemes. However, meiosis in these apomicts is an abbreviated version of meiosis in their sexual ancestors as it is asyndetic and restitutional. It is known that asyndesis in sexual meiotic mutants, and in the few sexual organisms that are naturally asyndetic (e.g. many male *Drosophila*), usually results from an abbreviation of Prophase I (Baker *et al.*, 1976), which is the longest stage of meiosis. For example, leptotene, zygotene and pachytene are absent from the meiosis of asyndetic male *Drosophila* (Lewin, 1980). It is

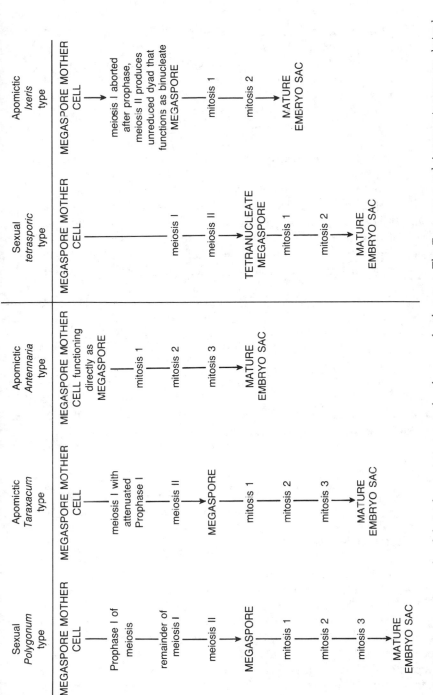

Figure 3.3 The ancestral sexual and derived apomictic types of embryo sac development. The *Taraxacum* and *Antennaria* types are derived from the *Polygonum* type, the *Ixeris* type from the tetrasporic *Drusa* or *Fritillaria* types. The sexual tetrasporic type is shown in generalized form to illustrate the number of divisions undertaken by *Drusa/Fritillaria* types. In fact, the divisions in the *Drusa* type result in an embryo sac that contains one haploid egg, two haploid synergids, two haploid polar nuclei and 11 haploid antipodals, whereas those in the *Fritillaria* type result in an embryo sac that contains one haploid egg, two haploid synergids, one haploid and one triploid polar nucleus and three triploid antipodals (further details are given in Willemse and Van Went, 1984).

reasonable to assume that Prophase I is similarly abbreviated in apomicts undergoing the *Taraxacum* scheme, so that the number of steps taken by these to produce an egg is reduced by 'a meiotic prophase' over the number taken by their sexual ancestors. Four divisions separated the generative cell from the egg in the sexual, tetrasporic ancestors of apomicts that exhibit the *Ixeris* scheme (Nogler, 1984). The first two of these comprised a reductional meiosis. The four nuclei of the resulting tetrad were enclosed within the same cell membrane, and this tetranucleate structure functioned as the megaspore. Each nucleus divided twice by mitosis to give the mature embryo sac. The *Ixeris* type of embryo sac development involves only three cycles of cell division. The two daughter nuclei formed by the first division remain enclosed within the same cell membrane and each undergoes two cycles of mitosis to give an eight-nucleate embryo sac. Thus, the number of steps taken to produce a mature egg is reduced by a whole division in these apomicts compared to the number taken by their sexual ancestors.

A similar reduction in the number of steps taken to produce a mature egg is found in aposporous apomicts. Here, a somatic cell within the ovule differentiates as a megaspore and undertakes two or three mitoses to produce a four- or eight-nucleate embryo sac. Thus apospory is similar to the *Antennaria* scheme in that the whole of meiosis is omitted. The maternal cell that functions as a mother cell is separated from the egg by only two or three cycles of cell division, instead of the five associated with sexual reproduction.

It is clear that both generative and aposporous apomicts undertake fewer steps in the production of eggs than did their sexual ancestors. It is reasonable to conclude from this that the amount of time required to produce a mature egg should be greater in a sexual ancestor than it is in an apomict. That is, acquiring the capacity to avoid meiotic reduction results in the capacity for precocious embryony or precocious oogenesis also being acquired. There are several strands of evidence that can be used to provide support for this hypothesis. The first relates to the incidence of the *Taraxacum* and *Antennaria* schemes. Many more generative apomicts follow the more abbreviated *Antennaria* scheme than follow the *Taraxacum* scheme. This is to be expected if precocity is a consequence of the pattern of avoidance of meiotic reduction, but it is not to be expected if it is a separately controlled phenomenon. Briefly, precocity must be sufficiently advanced to result in fertilization being avoided. If precocity is a consequence of the pattern of avoidance of reduction, it will be more advanced under the *Antennaria* scheme than under the *Taraxacum* scheme. Acquiring the *Antennaria* scheme is therefore more likely to result in fertilization being avoided, and is therefore more likely to result in a viable pattern of apomixis. Consequently, the majority of generative apomicts should exhibit the *Antennaria* scheme, as is the case. In contrast, if precocity is acquired separately, the pattern of avoidance of reduction should not affect the degree of precocity. The avoidance of fertilization should be as efficient in taxa acquiring the *Taraxacum* scheme as in those acquiring the *Antennaria* scheme and, contrary to what is found, each should be equally frequent. The phenomenon of facultative apomixis also provides some confirmation that precocity is a consequence of the pattern of avoidance of meiotic reduction. In facultative apomicts, unreduced eggs develop precociously but reduced eggs develop normally. This is to be expected, as only the unreduced eggs are produced by an abbreviated pathway. It is possible that this behaviour could result from a gene for precocity being expressed only in unreduced eggs, but in the absence of firm evidence that this is the case, it is unreasonable to favour this hypothesis over the more parsimonious one proposed here. Finally, strongly supportive evidence is provided by aberrants that have reverted from the apomictic pathways of egg production to ancestral pathways. Such aberrants have been observed in *Parthenium argentatum* (Powers, 1945) and *Taraxacum* (Sørensen and Gudjónsson, 1946; Sørensen, 1958). Apomicts of both taxa exhibit precocious embryony (Richards, 1970b; Esau, 1946). The aberrant *Parthenium* lineage reverted from the *Antennaria* scheme to the *Polygonum* scheme. The aberrant *Taraxacum* lineages reverted from the *Taraxacum* scheme to the *Polygonum* scheme. In both cases, the reductional meioses undertaken by the aberrants were disturbed and resulted in some restitution, but in both cases many of the unreduced eggs (as well as almost every reduced egg), were fertilized. (That the reductional meioses were disturbed is not surprising. The *Taraxacum* aberrants were hypotriploid (3n − 1) and the pairing of homologous and homoeologous chromosomes during meiosis I would have been irregular. In general, pairing of chromosomes in any apomictic lineage that suddenly reverts to a syndetic meiosis may be disturbed, simply because the previous absence of meiosis, or of syndesis, will have allowed the build up of numerous chromosome mutations.) The observation that is of most interest is not that meiosis was disturbed, but that many of the unreduced eggs that were generated by the ancestral, unabbreviated, *Polygonum* scheme were fertilized. This loss of precocity is only expected if precocity is a consequence of the pattern of the avoidance of reduction for it is highly unlikely that, if these capacities were separately controlled, they would be lost simultaneously.

The fact that there was a tendency in these aberrants for unreduced eggs to be fertilized demonstrates that the simple avoidance of meiotic reduction is not sufficient to avert fertilization. Rather, fertilization will be avoided only if the avoidance of reduction results from a process that also involves an acceleration of the pace of egg production and maturation. This is achieved by the *Antennaria* and *Taraxacum* schemes, but not by the *Polygonum* scheme undertaken by the aberrants. However, it is important to acknowledge that some unreduced eggs did develop parthenogenetically, even though they were generated by the *Polygonum* scheme. This tendency increased with the ploidy level. This was most noticeable in *P. argentatum*, where eggs produced by parents exhibiting very high ploidy levels were more likely to develop parthenogenetically than those produced by parents exhibiting lower levels. However, most eggs were fertilized, irrespective of their ploidy level. This indicates that the level of DNA within an egg can influence the probability of it developing parthenogenetically, though not sufficiently to provide organisms with an efficient system of asexual reproduction. The relationship between the level of DNA and parthenogenesis may be due to high levels of DNA inducing precocity. The accelerating effect of high levels of DNA on development are observed in a number of tissues (Pandey, 1955; Levin, 1983), although its effect on eggs has not been clearly determined.

3.4 SUMMARY

A number of points have been raised in this chapter, and it is worthwhile to summarize them before moving on to discuss other issues.

1. There is an intrinsic cost of sexual reproduction in anisogamous outbreeding animals and heterosporous plants. This cost does not result from a change in fecundity, as the avoidance of meiotic reduction does not result in a reduction in the number of meiospores or eggs produced. Rather, the cost results from the nature of the sexually reproduced offspring. Thus in dioecious taxa the cost of sex is that of producing males. In cosexual taxa it results because some of the sexually reproduced progeny are asexual, the eggs from which they developed having been fertilized by meiotically reduced male gametes produced by asexual forms – a proportion of these male gametes will have contained the gene or genes for asexuality. There is no cost of sexual reproduction if reproduction is by obligate selfing. In this situation, mutations for asexual reproduction are selectively neutral.

2. There is an intrinsic cost of asexual reproduction in anisogamous homosporous plants. This results partly from a 50% reduction in fecundity of the female function and partly because the male function of cosexual forms is largely unable to act as a carrier of genes

for asexuality as male gametes are meiotically unreduced. The capacity for asexual cosexuals to function as sexual male parents is therefore very restricted, as through this function they will give rise to inviable polyploid progeny. The reduction in the fecundity of the female function is due to the avoidance of meiotic reduction resulting in a halving of the number of meiospores produced.

3. Among the four major groups of land plants, obligate asexual reproduction is unevenly distributed between and among homosporous and heterosporous forms. Among homosporous forms, it is common in the pteridophytes but absent from the bryophytes. Among heterosporous forms it is common in the flowering plants but absent from the gymnosperms.

Because of the cost of asexuality associated with homospory, obligate asexual reproduction in homospores is selected as an alternative to female sterility, not as an alternative to sexual reproduction. Most female sterility in pteridophytes is probably a consequence of hybridization, not of mutation. Thus the incidence of obligate asexuality in these is positively correlated with that of hybridization. Hybridization is uncommon among bryophytes, and it is the absence of hybridization-induced female sterility that largely accounts for the absence of obligate asexuality among these. Female sterility will also arise through mutation. However, asexuality is less likely to become established when female sterility is generated by mutation than when it is generated by hybridization. This is because an asexual, female-sterile lineage will become established only if it is isolated from sexual relatives. A novel hybrid genome is much more likely than a non-hybrid genome to be equipped to occupy a new ecological niche that is unsuitable for its sexual progenitors.

Because of the cost of sex in heterosporous forms, obligate asexuality in these is selected as an alternative to sexual reproduction, not as an alternative to sterility. It is absent from gymnosperms but present in flowering plants because it is much more easily evolved by the latter than by the former. In flowering plants, the egg usually matures in synchrony with pollen maturation or release, so that the male gametes gain access to the eggs as the eggs become receptive to them. In asexual forms, the eggs mature before the pollen is matured or released. As a consequence, the eggs are able to realize their innate capacity for parthenogenesis. The early maturation of the eggs of asexual forms is a pleiotropic effect of the mutation that causes the megaspore to be meiotically unreduced. Thus only a single mutation – for the avoidance of meiotic reduction – is required to generate an asexual phenotype in flowering plants. In gymnosperms, pollination occurs before egg maturation, so that male gametes are in position to fertilize the eggs as soon as they mature. In order to evolve asexuality, gymnosperms would have to acquire a mutation to prevent pollination as well as one to prevent meiotic reduction. Moreover, these mutations would have to be acquired in such a way that they were expressed simultaneously, or almost so, as the expression of either one without the other would be highly deleterious, resulting in either the fertilization of unreduced eggs or in the production of haploid embryos. This constraint is a considerable one which no gymnosperm has been able to overcome.

4. During the evolution of the land plants it is probable that different cells within the axial row of the archegonium have functioned as the egg. This has had an important bearing on the capacity of different taxa to evolve parthenogenesis. The eggs of homosporous forms generally appear to be incapable of developing parthenogenetically, as their DNA is so highly dispersed that it requires a template – provided by the male gamete – to condense to a state compatible with mitosis. In flowering plants, the polar nucleus appears to be the homologue of the homosporous egg, whereas the functional egg is homologous with the ventral canal cell of the fern archegonium which retains mitotic potential. It is pertinent that many asexual flowering plants require the polar nuclei to be fertilized if they are to produce an endosperm. That is, the cell that, in flowering plants, is homologous to the homosporous egg is often incapable of dividing unless it has been fertilized. This suggests that if substitution of the egg had not occurred during the evolution of the flowering plants many taxa that have been able to evolve parthenogenesis would not have been able to do so. That is, the capacity of flowering plants to reproduce asexually is largely due to this group adopting as an egg a cell that is not homologous to the ancestral egg.

Chapter 4

Cosexuality, asexuality, and the male function

4.1 INTRODUCTION

Asexuality is a characteristic only of the female function. But most asexual taxa that are descended from cosexual sexual ancestors have retained the male function. Often, this function is only partly effective; for example, the male meiosis in many obligately asexual taxa is reductional but disturbed and results in a high proportion of the gametophytes/gametes produced being genetically unbalanced and thus inviable or otherwise ineffective (e.g. Weimarck, 1973). This is perhaps more likely to be a problem in anisoploid than in isoploid individuals. For example, Liljefors (1955) was unable to germinate pollen from triploid apomicts of *Sorbus* (whitebeams and rowans) but recorded 5–25% germination of pollen from tetraploid apomicts. Nevertheless, there are a number of instances of asexual individuals being used successfully as male parents, indicating that many individuals are able to produce at least some viable and effective male gametes. This is the case, for example, in the apomictic flowering plant taxa *Potentilla argentea* (Muntzing and Muntzing, 1945), *Pennisetum ciliare* (Cenchrus ciliaris; Taliaferro and Bashaw, 1966), *Paspalum notatum* (Burton and Forbes, 1960), *Panicum maximum* (Savidan, 1980), *Poa pratensis* and *Poa alpina* (Åkerberg and Bingefors, 1953), *Taraxcum* (Richards, 1970a, 1973; Morita *et al.*, 1990) and *Hieracium* (Ostenfeld, 1910).

It seems that many of the asexual taxa that are descended from cosexual sexual ancestors have amassed the benefits that may result from asexual reproduction via the female function while retaining a hold on sexuality via the male function. And for good reason. The male function is a precious asset. It was shown in Chapter 3 how it may play an important beneficial role during the establishment phase of asexuality in cosexual organisms, enabling the asexual component of a population to act as male parents in crosses with the sexual component. In so doing, copies of the gene or genes for asexual reproduction are introduced into the genomes of the progeny of sexual mothers. In situations where reproductive success via the female function is the same in sexual and asexual individuals this ability of asexual individuals to sire progeny tips the balance of selection in favour of the genes for asexuality, which are driven to fixation. However, the role of the male function in cosexual asexual organisms is not limited to helping asexuality to become established. It has other roles that will be explored in this chapter. It will be shown how it can improve the genetic background against which genes for asexual reproduction must operate and how under certain conditions it can provide the means by which asexual lineages can track changes in the environment. It will also be shown how under certain conditions it can overcome the cost of asexual reproduction in homosporous organisms and how it may play an important role in determining the geographic distribution of related sexual and asexual forms. But in addition to these beneficial effects it will be shown how the male function can be a burdensome accompaniment, reducing the reproductive success of a pseudogamously apomictic lineage through a spectacular act of treachery, by enhancing the reproductive success of neighbouring competing lineages.

Before moving on to discuss these issues in detail it is important to stress that the roles that will be ascribed to the male function can be fulfilled even when most of the male gametophytes or gametes produced by an individual may be inviable or otherwise ineffective. All that is needed is that some viable gametes are produced. Thus it is demonstrated in Chapter 3 how it may be necessary for apomictic flowering plants to sire, on average, only one progeny each in order to confer an overall advantage on a gene for asexual reproduction which arises in a cosexual sexual population. Clearly, the more effective the male function is then the more rapid or pronounced will be its effects. But even a minimally effective male function will often be sufficient to give asexual individuals the selective edge over related sexual forms when they confront each other in the ecological arena.

4.2 THE MALE FUNCTION AND THE FITNESS OF ASEXUAL LINEAGES

One aspect of the male function is that, in the presence of sexual compatible forms, it provides an asexual cosexual individual with the opportunity to produce asexual progeny that have fitnesses that are greater than its own. For example, in the model described by Expressions (3.4), inbred asexual parents with fitness $1 - d$ give rise via their male function to (sexually produced) asexual progeny with a fitness of 1. In contrast, these parents give rise via their female function to (asexually produced) asexual progeny which, because they receive the parental genome intact, have the parental fitness of $1 - d$. The mating regime within this model is closely circumscribed, with reductions in fitness being due only to inbreeding. However, the point made is a generally valid one. Typically, a population will comprise individuals which differ greatly in fitness for a number of different reasons – inbreeding depression will be just one such reason. It would be very fortuitous if an asexual mutation arose in the most fit individual. It is much more likely to arise in an individual with inferior fitness. However, a carrier of the mutation will, through its male function, sire some sexually reproduced progeny. Some of these will receive a copy of the mutation and will therefore be asexual, and some of these will have genotypes that are superior to that of the carrier. As the genomes of which this mutation is part are inherited in their entirety (barring mutations) via the female function, an asexual mutation arising in a cosexual population could find itself in an enviable position. Through the male function of its carriers it can move from genome to genome, infiltrating into that part of the population with the highest fitnesses, while through the female function of its carriers it can fix those genomes that generate particularly well-adapted phenotypes.

This ability to hop from genome to genome, sampling phenotypes, is a characteristic of genes for asexuality that arise in cosexual asexual taxa that express an at least minimally effective male function (i.e. which produce sufficient viable male gametophytes/gametes to be able to produce at least one progeny via this function). In these, the ultimate fate of a gene for asexuality is not tightly linked to that of the genome in which it arises. Such linkage will only occur if the gene for asexual reproduction results in the male function being rendered ineffective. But with an effective male function the gene can spread to different and often better genomes. This is in marked contrast to the fate of a gene for asexuality that arises in a dioecious taxon. Here, the fate of this gene is inextricably bound to that of the genome in which it arises, as it can only be propagated by the propagation of this genome. Thus the chances of a gene for asexuality succeeding, or of it finding residence in a well-adapted genome, may be higher if it arises in a cosexual taxon which exhibits an effective male function than if it arises in a dioecious taxon.

The role of the male function in providing the opportunity for a gene for asexuality to genome hop has important repercussions for the long-term maintenance of asexuality. This issue will be considered in some detail in Chapter 6, but it is pertinent to introduce it here. Briefly, there is a widely held opinion that the intrinsic cost of sexual reproduction compared with asexual reproduction (in animals and heterosporous plants) may be overcome in ecological situations where the advantage or disadvantage of possessing particular genes or gene combinations may change from generation to generation (e.g. Maynard Smith, 1978, 1980, 1988b; Bell, 1982; Michod and Levin, 1988). Such switches in the selective value of a gene or gene combination may be caused by a variety of factors. One phenomenon in particular has attracted the interests of biologists. This concerns the types of genetic changes that characterize host-pathogen or host-parasite interactions (e.g. Levin, 1975; Jaenike, 1978; Bremermann, 1980; Hamilton, 1980; Hamilton *et al.*, 1990). Here, a genetic strain of pathogen or parasite against which a host generation has little or no resistance will increase greatly in frequency. Sexual reproduction equips a host population to deal with this situation by providing the mechanism by which resistance genes can be rearranged into new effective combinations. Thus a sexual parent that is susceptible to a pathogen or parasite will produce genetically variable progeny, some of which may be resistant to the pathogen or parasite. In contrast, an asexual parent will produce progeny that are as susceptible as the parent to the pathogen or parasite. Consequently, the fitness of asexually produced progeny may be much less than that of sexually produced progeny. Whenever pathogen or parasite pressure is intense this difference in fitness may be sufficient to overcome the intrinsic cost of sex when compared with asexual reproduction, enabling sexual individuals to persist.

This argument is a powerful one when it is applied to dioecious asexual organisms. But it is less so when it is applied to cosexual asexual organisms that possess an effective male function. In the presence of sexual conspecifics, this function enables asexual forms to track changes in the environment. Through this function the asexual component of a population will be able to obtain new genes and new gene combinations; any genetic advantage the sexual component of a population may initially exhibit will be shared with the asexual component of the population. That is, an effective male function unites the gene pools of the sexual and asexual components of a population.

Nevertheless, this is one area where an effective male function may provide asexuality with only a temporary respite from severe ecological pressure. The asexual component of the population will be able to track changes in the environment only via its male function. It produces genetically invariable lineages through its female function. Consequently, its maternally generated progeny may, on average, have much lower fitnesses than its paternally generated progeny; the former may also have much lower fitnesses than both the maternally and paternally produced offspring of the sexual component of the population. Thus the continued survival of the asexual component of the population will be largely dependent on the extent to which its female function is made ineffective by its lack of ability to track environmental changes. To see this it is only necessary to refer back to the modified form of Expression (3.4b) which shows that a gene for asexuality will increase in frequency when

$$f + 0.5m > 1.$$

f is the quotient of (fitness of the asexual component of the population via the female function/fitness of the sexual component of the population via the female function) and m is the equivalent quotient for the male function. m is very unlikely to have a value greater than 1. Indeed, if the male meiosis is disturbed it will have a value much less than 1. Consequently, if asexuality is to persist in the population the value of f must approach 1. This will not be achieved if the success of an offspring is dependent on it receiving genes or gene combinations that were not present in its mother.

4.3 THE MALE FUNCTION AND THE MAINTENANCE OF ASEXUALITY

It is clear that the male function may play an important role in the spread to fixation of a gene for asexual reproduction, through its ability to impart on the carriers of this gene the capacity for sexual reproduction. By doing so, it increases the number of offspring its carrier produces and also provides a mechanism by which the gene for asexuality can infiltrate into the most well-adapted genomes.

The capacity for sexual reproduction imparted by the male function can also be a very useful attribute in the post-establishment maintenance phase. This is especially the case in homosporous organisms, where there is a potential twofold cost of asexuality. Superficially, the cost of asexuality in these is such that it would seem that asexuality will persist only if asexual populations remain totally isolated from related sexual populations. This is not the case. Although a sexual individual migrating into an asexual homosporous population will, in theory, have a considerable fitness advantage over the asexual natives, this may not be sufficient to guarantee that, in actuality, the capacity for sexual reproduction via the female function will spread to fixation in the population. Asexuality will be vigorously defended on two fronts, the manning of which shows a sexual invader to be the stumbling possessor of not one Achilles' Heel but two.

The first weakness demonstrated by an invading sexual homosporous individual concerns the method of invasion itself. This will be by a meiospore, which in most taxa will be air-borne. This is the only part of the invader that can be dispersed freely over large distances and is thus the only part of the invader that is likely to reach an asexual population. The meiospore will germinate to give a gametophyte, an egg of which must be fertilized if the invader is to achieve reproductive success. The problem here is that this egg will have access to only two sources of male gametes – those produced by its own gametophyte, resulting in selfing by intragametophytic non-parthenogenetic automixis, and those produced by neighbouring gametophytes. Unless the density of invading sexual meiospores is extremely high, the neighbouring gametophytes will have been produced by the resident asexual component of the population. They and their male gametes will be meiotically unreduced, and each gamete will carry all of the genetic information needed to generate an asexual phenotype.

If the invader's embryos are produced by selfing, they will be sexual but 100% homozygous and may suffer from severe inbreeding depression. If they result from outcrossing with asexual neighbours they will be at an enhanced ploidy level and they may be sexual, asexual or sterile. They will be asexual if the genetic control of asexual reproduction is dominant, sterile if it is not but if the ploidy level attained is incompatible with a balanced reductional meiosis. Thus the probability of sexuality surviving the first hurdle will be low, although it is important to note that if sexuality is lost the sexual immigrant's genes may nevertheless persist in the population, possibly enriching it, if the progeny it produces are viable but asexual. In a very real sense, the first defence of the members of an asexual population against invasion by sexual forms is not to repel these outright, but to counter-attack by invading the invader's genome. In this way, the sexual invader's genes can be plundered and put to good use, becoming incorporated into the genomes of the asexual lineages that provided the male parents.

Occasionally, the invader's progeny will be both sexual and viable. In this case, the second weakness of a sexual invader will be exposed. This is that the fitness advantage it exhibits over its asexual neighbours is manifested totally from its capacity to produce twice as many meiospores from each sporangium as these (Chapter 3). That is, the fitness advantage is not one that will help a sexually reproduced, sexual embryo to survive the juvenile stage and develop into a reproductively mature sporophyte. The resident asexual population will have been previously pruned by natural selection to fit the selective contours of its habitat. It is unlikely to be faced by an invader with a better-adapted phenotype. Consequently, many of the invaders will be repelled before they can utilize their twofold advantage to establish a bridgehead. Sexuality will once again have been defeated, but this time under rules of engagement which preclude the immigrants' genes from being incorporated into the pool of genes of the asexual component of the population.

Even if this second line of defence is breached, the asexual members of the population have reserve forces they can mobilize. Sexual reproduction will persist for as long as there is a supply of male gametes. Until the sexual immigrants achieve a high frequency in the population, the greater proportion of these gametes will be supplied by the asexual component of the population and will consequently be meiotically unreduced. Their utilization will drive the mean ploidy level of the sexual component ever upwards and will result in genes for asexuality accumulating in sexual lineages. Eventually, a ploidy level will be reached among the sexual component which either triggers a dosage response at the relevant loci such that sexually produced progeny adopt the asexual phenotype, or which results in sterile or inviable progeny.

Thus once an asexual homosporous population has become established it will be difficult to dislodge. Sexual forms that reach the population will be subjected to the insidious defensive tactic of genomic invasion, as their eggs are fertilized by male gametes generated by the asexual component of the population. The consequences of this include polyploidization and the donation of a full set of genes for asexual reproduction to each sexually produced individual. Although the genes of the invaders may persist in the population, it is very unlikely that the sexual phenotype will.

The maintenance of asexuality in populations of cosexual heterospores is not so problematic, as in this group there is a cost of sex. However, it is shown in Chapter 3 how the cost of sex may be wholly attributable to the activity of the male function of asexual forms. This will be the case whenever the transition from sexual reproduction to asexual reproduction does not result in asexual individuals achieving greater reproductive success than their sexual conspecificis via the female function. In this circumstance, if asexual individuals lack an effective male function, genes for sexual and for asexual reproduction may be selectively neutral with respect to each other. Consequently, an invading gene for sexual reproduction could achieve fixation through random genetic drift. I am not proposing that this is, or has been, a real threat to the maintenance of asexuality in this group. However, it is useful to draw attention to it, even if only to reinforce the point that much of the advantage of asexuality in cosexual heterosporous plants may be due to the presence of an effective male function. And even if the female function alone can impart an advantage to asexuality over sexuality, the possession of an effective male function will further extend this advantage.

4.4 THE MALE FUNCTION AND GEOGRAPHIC PARTHENOGENESIS

It has been clear for many years that related sexual and asexual forms often have different geographic distributions. In many taxa, sexual forms may be sympatric with asexual relatives, occupying an area in the centre of the latter's range, but the asexual forms extend over a greater area, and to higher latitudes and altitudes. This is the case in *Arnica* (Barker, 1966, cited in Bierzychudek, 1987b), *Eupatorium* (Sullivan, 1976), *Parthenium argentatum* (Rollins, 1949), *Poa* (Soreng, 1984), *Hieracium, Bothriochloa, Rubus* subg. *Eubatus* and *Antennaria* (Stebbins, 1971), and is commonly the case in *Crepis* (Babcock and Stebbins, 1938), *Townsendia* (Beaman, 1957) and *Taraxacum* (Richards, 1973). Bierzychudek (1987b) confirms the statistical significance of these trends in a review of the subject. In a number of cases, there is evidence that asexual forms are more able to occupy once-glaciated areas than related sexual forms (Porsild, 1965; Haskell, 1966; Bierzychudek, 1987b).

Numerous hypotheses have been proposed to explain these distributions. Levin (1975) proposes that asexuality is most likely to become established in areas where pressure from pests and pathogens is low. These areas will be away from the centre of distribution of the species and away from any other biological islands where sexual forms and their pests and pathogens may have concentrated (e.g. glacial refuges). As was shown in the previous section, this general negative association between asexuality and pest/pathogen pressure has attracted much interest; it is an aspect of a more general hypothesis that asexual organisms will tend to occupy environments where biological interactions are relatively unimportant (Glesener and Tilman, 1978; Bell, 1982). In contrast, Stebbins (1950), Bayer and Stebbins (1983), Haskell (1966) and Catling (1982) have argued that the wider distribution of apomicts may be due to their superior colonizing ability, as they, unlike their sexual relatives (who are usually outcrossers), are able to found a population with a single individual. More recently, Lynch (1984) has argued that selection among asexual forms has occurred for general-purpose genotypes that can persist under a wide variety of conditions. Such genotypes will enable these organisms to extend their range. Finally, a persistent argument has been that polyploidy, possibly associated with hybridization, will generate in asexual forms phenotypes that are absent from sexual relatives and that can cope with environments in which sexual forms would be unable to become established (e.g. Stebbins, 1971), thus enabling the asexual forms to occupy a range greater than that of their sexual relatives. Certainly, polyploidy is much more common among sexual forms, especially apomicts, than among their asexual relatives. And it is clear that polyploidy can have many effects on an organism, some of which will alter its ecological tolerance (Lewis, 1980). Bierzychudek (1987b) argues that the consequences of polyploidization are sufficient to account for the different distributions of related sexual and asexual forms.

However, none of these arguments considers the role the male function may play in determining the distribution of related sexual and asexual forms. And yet this role may be a very important one. Evidence that this is the case comes from *Taraxacum* (Mogie and Ford, 1988). This genus will be mentioned several times throughout the remainder of this book, and it is perhaps pertinent here to describe some basic aspects of its biology and history, as interactions between these and the male function may have helped to shape its current pattern of distribution.

Taraxacum comprises some 2000 extant species grouped into over 30 sections. Of these species 90% are obligately apomictic (reproducing by non-pseudogamous generative apomixis), and the remainder comprise facultative apomicts and obligate sexuals (Richards, 1973). The genus has a wide distribution, its European representatives being distributed from the Mediterranean to the Arctic. All of the apomicts (both obligate and facultative) are polyploid, typically triploid or tetraploid, but all obligately sexual forms are diploid. Sexual taxa have a wide geographic distribution, but even so extend over only about half of the area occupied by apomictic taxa (Richards, 1986). In Europe, they are concentrated in the southern part of the range of the genus, in areas that remained largely unglaciated during the Pleistocene glaciations (Tschermak-Woess, 1949; Fürnkranz, 1960, 1961, 1966; Richards, 1973; den Nijs and Sterk, 1980). They are sympatric with apomictic taxa over most of this area, although the latter are also common throughout northern Europe, where sexual forms are scattered and rare. A problem in interpreting this pattern of distribution is that, although the restriction of sexual forms to the more southerly parts of the range of the genus may indicate that they lack the ecological breadth of apomictic forms, within the areas where sexual forms are common they appear to be as ecologically robust as sympatric apomictic forms, occurring in the same range of habitats and at high and low altitude (den Nijs and Sterk, 1980). This weakens the argument that the different distributions of sexual and apomictic forms are primarily or mostly due to differences in ecology. An alternative explanation, centred around the male function of the apomicts, emerges if the history of the genus is considered. The following brief account of this history is a summary of Richards' (1973) more detailed study.

The genus probably originated in the western Himalayas during the Cretaceous. The original species were diploid and sexual and displayed karyological and morphological features that are described as 'primitive'. Approximately 50 primitive species are extant. Most are diploid and sexual and they persist, largely as relict populations, in montane areas in west and central Asia and the Mediterranean region. Interestingly, a number of these taxa are high alpines – usually snow-patch plants – of glaciated areas which are largely devoid of apomicts (Richards, personal communication). Polyploidy and apomixis arose early in the history of the genus, as did changes in gross karyological and morphological features. Two groups of derived plants can be recognized – precursor species and advanced species. Precursor species display a mixture of primitive and derived characteristics and are mostly diploid and sexual. Several are extant and occupy much the same range as extant primitive species. Advanced species display mostly derived characteristics. They are typically polyploid and apomictic, although there are some sexual, diploid forms. Advanced species make up the bulk of extant species of *Taraxacum*, and are thought to be responsible for extending the range of the genus into arctic and alpine areas. The major expansion of the genus was concluded by the early Pleistocene.

The history of *Taraxacum* subsequent to the period of major expansion has been greatly influenced by the Pleistocene glaciations. Richards (1973) considers that the mainly apomictic arctic-alpine taxa were pushed southwards during the glaciations. With the retreat of the glaciers, primitive and precursor species, which were mainly sexual and which had survived as isolated populations in glacial refuges, converged to produce a hybrid swarm. This migrated northwards and met the arctic-alpine taxa. Hybridization between these (with the arctic-alpine apomicts acting as pollen-donors) and between their immediate derivatives resulted in apomixis genes being transferred and in a large number of hybrids becoming fixed as apomicts. The new sexual and apomictic hybrids, which exhibited advanced characteristics, subsequently predominated in the cool-temperate zones between the arctic-alpine and warm-temperate areas colonized by their parental types.

Although massive migrations of apomictic and sexual types have undoubtedly occurred during the Pleistocene as a result of glaciation, these do not adequately explain the present distributions of advanced and primitive types and of sexual and apomictic types (Mogie and Ford, 1988). A major problem concerns variation in the distributions of advanced, precursor and primitive sexual forms. Advanced sexual forms are predominantly found only south of the area reached by the southern edge of the ice sheets formed during the most recent (Wurm) glaciation. However, primitive and precursor sexual forms predominate in an even more southerly location. This cannot be due to primitive/precursor sexuals being less able to withstand low temperatures than advanced sexuals; nor can the more restricted distribution of advanced sexuals relative to that of advanced apomicts be due to the sexual forms being less adapted to low temperatures than the apomicts, as many sexual forms are high alpines or at least occupy montane habitats. Nor can these distributions be due to sexual types lacking the competitive ability of apomicts, as both co-occur in habitats where competition would be expected to be intense (den Nijs and Sterk, 1980).

These problems can be overcome if consideration is given to the male function and the pattern of inheritance of apomixis in this genus. Male meiosis in *Taraxacum* apomicts is often reductional though disturbed. These disturbances, which appear primarily to result from the hybrid and often anisoploid nature of the genome, typically result in pollen exhibiting a range of ploidy levels from fully reduced to fully unreduced. Pollen at or near a euploid level appears to be able to fertilize eggs. Thus in a cross between a diploid sexual (acting as the maternal parent) and a triploid apomict, progeny at or around the eudiploid, eutriploid and eutetraploid level will be produced. In such crosses, diploid progeny are invariably sexual whereas polyploid progeny can be sexual but most are apomictic (Richards, 1970a,b, 1973; Morita *et al.*, 1990). These experiments have an important bearing on the nature of the genetic control of apomixis, as they indicate that an apomictic phenotype is inherited only if the apomict donates more than a haploid complement of genes to its progeny when it is used as a male parent in a cross with a sexual female. A similar situation also appears to exist in *Ranunculus auricomus* (Nogler, 1984). These observations will be returned to in Chapter 5. For the moment, I only wish to draw attention to the fact that they pose problems for the maintenance of sexuality in an expanding population of mixed sexuals and apomicts.

Taraxacum seeds are enclosed within single-seeded fruits (achenes) which are individually wind-dispersed. Most, however, fall close to the mother plant (Mogie and Ford, 1988). Consequently, there will be two types of migration at the end of a glacial period as previously glaciated areas become available for colonization. The first will be a gradual (few metres per year) invasion of new territory, as achenes from plants at the edge of the population are deposited at high density into immediately adjacent uncolonized sites. This type of migration will have enabled populations that persisted in glacial refuges during the last glacial period to have expanded their range by, at most, a few tens of kilometres during the present interglacial period. The second type of migration will involve colonization of sites distant to the parental population by the long-range dispersal of a small proportion of the achenes. This will result in individuals becoming established at low density in 'founder populations' in front of the main body of advancing migrants. Most sexual taxa are self-incompatible. Pollination is insect mediated. Consequently, isolated sexual individuals in newly established founder populations will receive little or no pollen and thus set few or no seeds. Seed set in isolated apomicts (which are non-pseudogamous) will not be pollen limited. Thus if these founder populations begin to increase in size through the local dispersal of achenes, the proportion of apomicts in the population will increase. Thus, as founder populations increase in size they will be increasingly dominated by apomicts. As population size increases pollen will become less limiting. Consequently, any sexual lineages that have persisted in the population will experience greater reproductive success. But because pollination in *Taraxacum* is insect mediated sexual individuals will receive most of their pollen from their nearest neighbours which, increasingly, will be apomicts. These crosses will result in a high proportion of the progeny of the sexual mothers being apomictic, causing an even greater decline in the proportion of sexual types in the population. Eventually, sexual forms will disappear, and the northern front of the range of the genus will be occupied by apomictic or predominantly apomictic populations. This front will form an impenetrable barrier to migration northwards, or from glacial refuges in general, of sexuality, as whenever a sexual individual invades this front it will be pollinated by apomicts, and its progeny will be hybrid apomicts. In effect, sexuality will be unable to spread northwards, and will be largely restricted to latitudes sexual forms occupied during the periods of glaciation.

This last point is important as it indicates why sexual forms are restricted to the south of the range of the genus and why primitive/precursor sexuals are concentrated even further south than advanced sexuals. The primitive/precursor sexuals reached Europe early in the Pleistocene and must have persisted in southern refuges through several ice ages. Some of the previous glaciations were more severe and extensive than the most recent and would have pushed the vegetation zones much further south (John, 1977). Thus the Saalian/Riss glaciation was more extensive and severe than the most recent Weichselian/Wurm glaciation. Advanced sexual taxa formed in the interglacial between the Riss and the Wurm would have been pushed south to a lesser extent than primitive and precursor sexual taxa established before the Riss, and would be expected to be found further north than these types.

This account of the causes of the restricted distribution of sexual forms and the more extensive distribution of asexual forms in *Taraxacum* is attractive because it does not have to resort to the assumption that asexual forms are more widely dispersed because they are in some way ecologically superior to sexual forms. The restriction of most sexual taxa to southern parts of the range of the genus, or to glacial refuges in other parts of the range, occurs because the patterns of rapid, long-distance migration required to disperse these taxa away from their glacial refuges will result in the loss of sexuality and of species identity as these migrants hybridize with resident apomicts. However, it need not result in the loss of the genes of the migrants, as these will be incorporated into hybrid apomictic genomes.

There is no reason to assume, and no evidence to indicate, that the ecology of sexual and asexual forms of *Taraxacum*, or the control of apomixis in this genus, is radically different from that of other cosexual agamic complexes. Consequently, the explanation proffered for the distribution of the *Taraxacum* agamic complex may be suitably applied to other complexes which show a similar tendency for asexual forms to be more widely distributed than sexual forms. That is, the activity of the male function may have had a considerable effect on the geographical distribution of sexual and asexual forms in several agamic complexes.

This is not to say that the distribution of related sexual and asexual forms cannot be influenced, perhaps significantly so, by other factors. Thus polyploidy can have many effects on an organism, some of which may alter its ecological tolerance in a way that will adapt it to disturbed areas, including those disturbed by glaciation (Stebbins, 1971; Lewis, 1980). For example, it can affect the length of the mitotic cycle, growth rate, enzyme activity, seed size, seedling vigour, flowering time and duration, resistance to pathogens and pests, temperature optima for various physiological processes, and competitive ability (Van't Hof and Sparrow, 1963; Van't Hof, 1965; Baker, 1965; Bennett, 1972; Bennett and Smith, 1972; Levin, 1983). The problem is to determine the relative importance of these different factors. Certainly, there are several polyploid complexes (comprising diploid species/

races and descendant polyploid species/races) among sexual taxa where the polyploid forms have a much wider distribution than the diploid forms (see Table 6.1 of Stebbins, 1971). This lends support to the argument that polyploidization, which is often accompanied by hybridization, can greatly extend the range of a taxon by affecting the ecological amplitudes of affected individuals. However, polyploidization in sexual taxa is associated with problems that are analogous to those associated with the divergence of asexual forms from sexual forms in apomictic taxa.

A common route for the formation of a sexual polyploid appears to be by chromosome doubling of a sterile diploid hybrid. In many cases, the polyploid will be self-compatible even if the parental diploids are self-incompatible, as polyploidy can disrupt the balance between self-incompatibility alleles. This is especially the case if self-incompatibility is gametophytically controlled and if the polyploid is non-disomic at the relevant locus (Richards, 1986). Richards (1986) makes the point that self-compatibility will often be a requirement if polyploidy is to become established, as polyploids are likely to arise in situations where their nearest neighbours, and most likely mates, are diploid. A cross between a tetraploid and a diploid will generate sterile triploid progeny. Viable progeny will result only if the tetraploid self-fertilizes. This will allow polyploidy to become established within a diploid population. However, the acquisition of self-compatibility has a further consequence as it can help to extend the range of the taxon or at least to accelerate the rate of expansion of the taxon into new areas, as the self-compatible forms can individually found populations outside the current range of the self-incompatible forms. Self-incompatible forms are much less efficient at founding an isolated population as this will require two seeds to fall close enough together to allow cross-pollination to occur. This advantage of self-compatibility is independent of ploidy level – it has long been recognized (e.g. Baker, 1955, 1966, 1967) that ecologically marginal or peripheral populations of many taxa are more likely to be self-fertile than those that are non-marginal or more central, even if there is no difference in ploidy between the marginal and more central populations (Ernst, 1953; Davies and Young, 1966; Vasek, 1968; Grant, 1975; Levin, 1975; Lloyd, 1980).

It is therefore possible that when a largely self-incompatible sexual taxon whose range has been restricted, for example by glaciation, begins to expand its range, the expansion may be spear-headed by self-compatible forms. If the taxon is involved in hybridization and polyploidization, polyploids will be over-represented in the self-compatible component and consequently in the spearhead. The diploid self-incompatible component could consequently become surrounded

by marginal polyploid self-compatible populations. This will greatly
retard the rate of expansion of diploid forms. As these advance on the
marginal polyploid populations they will cross with polyploid forms
to produce sterile, anisploid progeny (the polyploids, being sexual, will
be isoploid). The polyploids will be able to produce viable, isoploid,
non-sterile progeny by self-fertilization. Thus the polyploids will form
an impenetrable barrier within the confines of which the diploid component of the polyploid complex will be restricted. However, the polyploids will continue to expand their range because of their ability to
self-fertilize. Self-incompatibility could then become established in
polyploid populations residing in the lee of the advancing front of
polyploids. This would not retard the expansion of the polyploid component of the complex as long as self-compatibility remained a characteristic of the advancing front.

This scenario is speculative. However, it is no more so than the
more traditional view that the greater range occupied by the polyploid
component of a sexual polyploid complex is primarily due to polyploidy conferring greater ecological vigour or tolerance. It is mentioned
here because this traditional view has directly influenced the interpretation of the distribution of the sexual and asexual components of
agamic complexes, as polyploidy is typical among the asexual component but is not nearly so common among the sexual component. The
point is that the ecological consequences of polyploidy in both sexual
and asexual polyploid complexes may have been exaggerated. In both
cases, the range of the diploid component of the complex (which will
also be the sexual component in agamic complexes) may have been
restricted primarily because of the breeding system of the polyploid
component (either self-compatible or apomictic) rather than because
the diploid component is ecologically inferior to the polyploid component over much of the complex's range. The polyploid component is
able to migrate much faster than the diploid component simply
because it lacks a requirement for mates. By migrating more rapidly,
it can erect a barrier to further migration of the diploid component,
this barrier involving either the imposition of the apomictic phenotype
on the progeny of the diploid sexual component (in agamic complexes),
or (in sexual complexes) the production of sterile, anisoploid progeny
by the diploid component when it crosses with the polyploid component.

4.5 THE MALE FUNCTION AND PSEUDOGAMY

So far in this chapter the male function has been depicted as a very
useful attribute of asexual organisms. Each of the roles described is

the result of individual apomicts achieving enhanced fitness through their male function. However, the male function plays a much more complex role in pseudogamous apomicts, once this system of reproduction has reached fixation in a population (clearly before fixation it will aid the spread of pseudogamous apomixis in much the same way as it will aid the spread of apomixis in autonomously apomictic taxa). Pseudogamy is particularly common among aposporous apomicts (Richards, 1986). Most pseudogamous apomicts are descended from outbreeders (Gustafsson, 1946–47) and the pollination syndrome this entails appears to have been retained by them. Consequently, if a pseudogamously apomictic individual is unable to use its male function to fertilize its endosperms this function will have no direct effect on its fitness. However, it can have an indirect effect by influencing the fitness of the individual's pseudogamously apomictic neighbours by fertilizing their endosperms. The significance of this is best appreciated by comparing seed development in autonomous and pseudogamous apomicts.

The endosperm in autonomous apomicts is produced by the fusion of female gametophyte nuclei. As these nuclei are all genetically identical to each other, being mitotic derivatives of the same megaspore, all of the genetically identical embryos produced parthenogenetically by a plant will be nurtured by endosperms that are also genetically identical to one another. The success of an autonomous apomict will therefore be dependent both on its ability to produce innately viable embryos and on its ability to produce, unaided, efficient endosperms. In contrast, although all the embryos produced parthenogenetically by a pseudogamous apomict are genetically identical, the endosperms nurturing these will be genetically variable, due to the presence of the genome donated by the pollen donor. Thus the efficiency of endosperms could vary. In effect, the fitness of a pseudogamously apomictic mother will be influenced by the quality of the pollen donor's genes, as well as by the quality of her own genes, in much the same way as the fitness of a sexual female is. Nogler (1984) states that seed-set in pseudogamous apomicts can be greatly influenced by the male pollen donor. Thus pseudogamous apomicts can exhibit a kind of seed incompatibility, especially if the pollen donor is at a different ploidy level to the maternal plant. The causes of this are not clear, although there is some evidence that a genetically unbalanced endosperm will degenerate. Nogler (1984) also describes the phenomenon of pseudogamous heterosis, in which the vigour of the asexually produced offspring is dependent on the identity of the pollen donor. This phenomenon has been observed in *Rubus* (Haskell, 1960) and *Malus* (Schmidt, 1964). Souciet (1978) reports that seed dormancy and rates of germination are affected by the identity of the pollen donor in *Panicum maximum.* In conclusion, it is becoming clear that the fitness of genetically identical asexually produced embryos may vary in pseudogamous apomicts due to the variation in the genetic composition of the endosperm that results from the incorporation of the genome of the male gamete into the endosperm initial.

Whereas it has been clear for some time that the pollen donor can affect the fitness of the recipient in pseudogamously apomictic taxa the issue of whether or not the pollen recipient can influence the fitness of the donor, as a result of receiving pollen, has received scant attention. Maynard Smith (1978: 66) argues that the fitness of the pollen donor can be enhanced by it donating pollen because

> . . . individual plants will tend to grow close to their relatives – i.e. to members of their own clone. Hence by producing pollen a plant is helping to ensure the fertility of plants genetically identical to itself; pollen production will be favoured by kin selection.

However, it is by no means clear that this will be the case. Populations of apomicts can comprise many different lineages. For example, Richards (1986: 407) points out that for populations of related apomicts (agamospecies)

> . . . for instance belonging to the same genus, many genotypes (agamospecies) may coexist. To take one example, H. Øllgaard (personal communication) has identified over 100 *Taraxacum* agamospecies inhabiting one hectare of waste ground near Viborg, Denmark. In the UK, it is commonplace to find 20–30 *Taraxacum* agamospecies coexisting. Such levels of diversity can also be found in some other agamic complexes (e.g. *Rubus* and *Ranunculus auricomus*), but in others (e.g. *Alchemilla* and *Sorbus*) it is unusual to find more than two agamospecies coexisting.

Taraxacum apomicts are non-pseudogamous, but the presence or absence of pseudogamy should not affect the pattern of seed dispersal. The point is that there is little evidence that the donors and recipients of pollen in pseudogamously apomictic taxa are usually from the same lineage. Another noteworthy point is that information is lacking on the extent to which pseudogamous apomicts are self-incompatible. Many appear to be descended from self-incompatible sexual ancestors, but the evolution of apomixis is often associated with polyploidization, which as has been shown can often result in the break down of incompatibility barriers. But in pseudogamously apomictic taxa in which self-incompatibility has been maintained, pollen donors and recipients will have to be from different lineages. Thus for various reasons it is by no means clear that the continued production of pollen by pseudogamous apomicts has been favoured by kin selection.

Many pseudogamous apomicts are also facultatively apomictic (Richards, 1986). In these, pollen production will be maintained because an individual's male function can enhance its fitness directly by providing it with the means to sire progeny. However, the issue to be considered here is whether or not pollen production can benefit the producer in obligately apomictic pseudogamous taxa. A useful way to approach this issue is to contrast the costs and benefits accruing to a pollen donor in a pseudogamous, obligately apomictic taxon with those accruing to a pollen donor in an autonomous, facultatively apomictic taxon. Pseudogamous apomicts require pollen for endosperm formation, non-pseudogamous facultative apomicts for sexual reproduction. This difference between the uses to which the male gamete is put is an important one. In facultative apomixis, the pollen donor's genes are incorporated into the embryo: the male gamete is a vehicle through which its producer can pass on copies of its genes to progeny. The fitness of the pollen donor will be positively correlated with that of the recipient. The male gamete does not serve this function in pseudogamy, as its genes are incorporated only into the genome of the sterile endosperm, not into an embryo (which is initiated parthenogenetically). Thus, whereas the male function of a facultative apomict provides it with the means to sire progeny, that of the pseudogamous apomict typically provides it only with the means to aid its pollen recipients, with which it may frequently be unrelated, to achieve reproductive success via parthenogenesis through their female function. This is not in the pseudogamous apomict's best interests. Transfer of pollen will predominantly occur between close neighbours, and close neighbours are likely to produce seed shadows that overlap to a considerable extent. Consequently, a pseudogamous apomict, via its male function, will be instrumental in helping its neighbour achieve reproductive success, but the price it pays for this misplaced altruism may be that its neighbour's progeny compete with the progeny it produces via its female function. Overall, the fitness of the pollen donor will either not be correlated with that of the recipient or will be negatively correlated. In this situation, a gene for male sterility may well be favourably selected and, indeed, male sterile strains have been observed in pseudogamous *Poa arctica* subsp. *caespitosa* and *Potentilla tabernaemontani* (Smith, 1963). Note that a male-sterile mutant should be viewed as something of a cheat in this situation as its reproductive success is dependent on its neighbours continuing to supply it with male gametes. Thus it is pertinent to ask whether a gene for male sterility will reach an equilibrium if it arises in a heterogeneous, pseudogamously apomictic population.

This problem will be approached by considering an obligately asexual pseudogamous population comprising n_2 cosexuals and a small number, n_1, of male sterile 'females'. The acquisition of male-sterility could have two consequences. The first is that resources that were previously allocated to the male function could be re-allocated to the female function, resulting in females having a higher relative fitness than cosexuals. The second consequence is that the pollen:ovule ratio in the population could decrease as the proportion of females increases, resulting in seed set being pollen limited. Let the relative fitnesses of females and cosexuals in the absence of pollen shortage be 1 and $1 - s$ respectively $(0 < s < 1)$. Their relative fitnesses in the presence of pollen shortage will be determined by whether or not they experience the effects of shortage equally. This is unlikely. For example, if a cosexual individual matures x ovules and a female $x + y$ ovules $(y > 1)$, and if each obtains only x effective pollen grains then mature seeds will be produced by all of the cosexual's ovules but by only a proportion $x/(x + y)$ of the female's ovules. Thus the female is experiencing pollen shortage but the cosexual is not. Overall, it is likely that for any given level of pollen shortage, the effects on the fitnesses of the two types will be asymmetrical. Consequently, in the presence of pollen shortage, the relative fitnesses of females and cosexuals will be assumed to be $1 - b$ and $1 - c$ respectively $(0 < b,c < 1)$. Note that $(s \leqslant c < b)$. It is unlikely that the number of pollen grains an individual receives will be primarily determined by the total availability of pollen in the population. That is, the number received will not be strongly positively correlated with the pollen:ovule ratio. Such a correlation is only to be expected if pollination is random. However, pollen recipients will receive most of their pollen from their nearest neighbours. Because of this, a simplifying assumption will be made – that pollen shortage is determined by the identity of the nearest neighbour, not by the pollen:ovule ratio. If this neighbour is female, the recipient experiences pollen shortage (i.e. recipient females have a fitness of $1 - b$ and recipient cosexuals of $1 - c$). If it is cosexual, there is abundant pollen (i.e. recipient females have a fitness of 1 and recipient cosexuals of $1 - s$).

A proportion $an_1/(n_1 + n_2)$ of females will have another female as the nearest neighbour, with the remainder having cosexuals as the nearest neighbour. Here, $n_1/(n_1 + n_2)$ is the proportion expected if dispersal is random and a represents a deviation from randomness. Females will typically be clumped, as seed-dispersal is non-random, so $a > 1$. Consequently,

$$\text{mean fitness of females} = \frac{an_1(1 - b)}{n_1 + n_2} + 1 - \frac{an_1}{n_1 + n_2}$$

or

$$1 - \frac{ban_1}{n_1 + n_2} \tag{4.1a}$$

The number of cosexuals with a female as the nearest neighbour is equal to the number of females that do not have other females as nearest neighbours. This is

$$n_1(1 - an_1/(n_1 + n_2))$$

The number of cosexuals with cosexuals as nearest neighbours is consequently

$$n_2 - n_1(1 - an_1/(n_1 + n_2))$$

The mean fitness of cosexuals is

$$\frac{n_1(1 - an_1/(n_1 + n_2))(1 - c) + (n_2 - n_1 + (an_1^2/(n_1 + n_2))(1 - s)}{n_2}$$

Rearrangment gives

$$\frac{n_2 - cn_1 - sn_2 + sn_1}{n_2} + \frac{can_1^2 - san_1^2}{n_2(n_1 + n_2)}$$

and finally,

$$1 - s - \frac{n_1(n_1 + n_z - an_1)(c - s)}{n_2(n_1 + n_2)} \tag{4.1b}$$

Equilibrium will be reached when Equations (4.1a) and (4.1b) are equal. Thus

$$1 - \frac{ban_1}{n_1 + n_2} = 1 - s - \frac{n_1(n_1 + n_2 - an_1)(c - s)}{n_2(n_1 + n_2)}$$

Rearrangement gives

$$ban_1 = s(n_1 + n_2) + \frac{n_1(n_1 + n_2 - an_1)(c - s)}{n_2}$$

$$ba = \frac{sn_2}{n_1} + c + \frac{n_1(1 - a)(c - s)}{n_2}$$

$$\frac{n_2}{n_1} = \frac{ba - c}{s} - \frac{n_1(1 - a)(c - s)}{sn_2}$$

and finally

$$\frac{n_2{}^2}{n_1{}^2} = \frac{n_2}{n_1} \cdot \frac{(ba - c)}{s} - \frac{c(1 - a)}{s} + 1 - a \qquad (4.1c)$$

Often $c/s \rightleftharpoons 1$. In this situation Equation (4.1c) approximates to

$$\frac{n_2}{n_1} = \frac{ab - c}{s} \text{ (approx.)} \qquad (4.1d)$$

When $a \rightleftharpoons 1$ this reduces to

$$\frac{n_2}{n_1} = \frac{b - c}{s}$$

In words, this is

<u>advantage to cosexual when there is pollen shortage</u>

disadvantage to cosexual when there is no pollen shortage

It is clear from Equation (4.1d) that mutations for male sterility could reach appreciable frequencies at equilibrium. For example, setting $s = 0.05$, $c = 0.1$, $b = 0.14$ and $a = 2$, the ratio cosexuals:females at equilibrium is 3.6: a population of 1150 pseudogamous apomicts will comprise 900 cosexual individuals and 250 male-sterile 'females'. It is thus surprising that male sterility has not been widely reported. However, the male function may be rendered ineffective in a number of ways: mutants that fail to produce anthers or pollen will be much more easily identified as male-sterile than mutants that simply produce inviable pollen, and yet the latter may be much more frequent than the former. It is thus of some interest that Nogler (1984) describes how male meiosis in pseudogamous apomicts is only rarely completely normal. He states that (p. 494) '. . . Obviously, pollen with a more or less reduced capability to germinate is sufficient to fertilize polar nuclei, whereby a considerable proportion of aneuploid pollen is tolerated.' However, it is worth considering the possibility that the observed reduction in pollen viability may be due to the selection of genes that reduce the effectiveness of the male function. The fitness of the carriers of these genes may be enhanced as a consequence, as their capacity to enhance the reproductive success of their neighbours by fertilizing their endosperms will be reduced.

Chapter 5

The genetic control of apomixis

5.1 INTRODUCTION

In Chapters 1–4, attention has been focused on the patterns of asexual reproduction exhibited by bryophytes and tracheophytes. The role that phylogeny and ontogeny have played in determining these patterns, the innate costs and benefits of sexual and asexual reproduction and the role played by the male function in cosexual forms in determining the spread, maintenance and geographic distribution of asexual reproduction have been described. In this chapter, I wish to move away from these matters and to concentrate instead on the genetic control of asexual reproduction.

As discussed in Chapter 1, most of the investigations into this subject have concentrated on understanding the control of aposporous and diplosporous forms of parthenogenetic apomixis in flowering plants. Other methods of asexual reproduction in this and other groups have received much less attention. This chapter must necessarily reflect this bias, as the validity of the arguments and models presented will be tested against published data. The arguments and models that will be presented are also relevant to understanding the control of apomixis in animals, as this reproductive process is equivalent to diplospory in flowering plants. This similarity between flowering plants and animals is apparent in all but name. Whereas parthenogenetic apomixis in animals is simply described as 'apomixis', this process in flowering plants is subdivided into 'diplosporous' and 'aposporous' forms. The term 'diplosporous' cannot be suitably applied to animals as they do not produce spores. Consequently, I have suggested the term 'generative apomixis' to describe collectively apomixis in animals and diplosporous apomixis in flowering plants (Mogie, 1988), as in both groups, but in contrast to the situation in aposporous apomicts, the parthenogenetically developing egg is derived from the generative cell (the oocyte or megaspore mother cell). I will use this term throughout this chapter.

The chapter will move carefully and gradually towards the construction of a model for the control of apomixis. A brief historical perspective will first be offered. The ground rules for the development of the model will then be laid and the model will be constructed. Its validity will then be tested against published data.

5.2   CONCEPTS AND CONUNDRUMS

The nature of the control of asexual reproduction, especially parthenogenetic forms, has attracted considerable attention for most of this century. One view is that asexual reproduction is an innate capacity of sexual taxa which usually lies dormant but which can be awakened by factors other than mutation. There is some supporting evidence for this view. Many sexual organisms exhibit tendencies towards certain aspects of asexual reproduction. For example, tychoparthenogenesis (the occasional production of embryos parthenogenetically by otherwise sexual organisms) has been observed in a number of animals and flowering plants (for reviews see Gates and Goodwin, 1930; Ivanov, 1938; Kostoff, 1942; Kehr, 1951; Kimber and Riley, 1963; White, 1973; Cuellar, 1974; Bell, 1982). Kimber and Riley (1963) record its presence in 71 flowering plant species distributed through 39 genera and 16 families, and its existence in insects has led to claims that it may be the first step in these acquiring asexual reproduction (e.g. Carson, 1961). Interestingly, a tendency for parthenogenesis can be greatly increased by artificial selection. This is the case in *Drosophila* (Stalker, 1954); Carson (1967) was able to select for an increase in the frequency of hatching of unfertilized eggs of *D. mercatorum* from 0.01 to 0.06. An increase in its frequency can also be obtained by manipulating the environment. Hakansson (1943) has demonstrated this in *Poa alpina*, which contains both apomictic and sexual biotypes; its frequency in sexual biotypes could be increased by delaying fertilization. This is an interesting result. In effect, this experiment generated conditions in which the egg was allowed to mature precociously with respect to pollination, enabling an innate tendency for parthenogenesis to be realized. This lends support to the proposal made in Chapter 3 that precocious oogenesis/embryony in apomicts results from the early maturation of the egg and is sufficient to explain parthenogenesis in many taxa.

In addition to many sexual organisms spontaneously exhibiting parthenogenesis, many can be induced to exhibit it by subjecting them or their eggs to a variety of stimuli, including temperature shocks, changes in pH, physical manipulation, delayed pollination or pollination with irradiated pollen (Kimber and Riley, 1963; Went, 1982). Bell (1982: 345) makes the important observation that there is likely to be substantial genetic variance for parthenogenesis in taxa in which it can be readily induced, as such variance is typical for any character that can readily be altered by manipulating the environment.

Tychoparthenogenesis and artificially induced parthenogenesis usually involve the development of an embryo from a meiotically reduced egg. The embryo is typically inviable in mammals (Beatty, 1967; Graham, 1974) and is frequently, but by no means always, so in insects (Went, 1982) and plants (Kimber and Riley, 1963). This raises the possibility that many examples of tychoparthenogenesis remain to be discovered, as it is unlikely to be stumbled across accidently in taxa where the embryo degenerates very early in development. Certainly, the immediate fate of unfertilized eggs of sexual flowering plants is not known for most taxa, although it is clear that in some (e.g. *Brassica napus*, Pechan, 1988), such eggs may occasionally divide a few times before degenerating. When viable, haploid plants are characterized morphologically by a reduction in size of all vegetative and floral organs (Kimber and Riley, 1963).

Although parthenogenesis is only one of several major components of asexual reproduction, the phenomena described above have prompted speculation that many organisms possess latent tendencies for asexual reproduction that can be realized under certain conditions. Two factors that have been identified as candidates for stimulating these latent tendencies are hybridization and polyploidy (e.g. Gustafsson, 1948). There is some evidence that these can have the proposed effect. Yudin (1970) found that colchicine-induced tetraploids of diploid maize lines produced significantly more embryos parthenogenetically from meiotically reduced eggs than their diploid analogues. Similarly, Quarin and Hanna (1980) generated facultatively aposporous tetraploids of *Paspalum hexastachyum* by treating sexual diploids with colchicine. Chromosome doubling has also been implicated in the expression of asexual phenotypes in *Dichanthium* (Reddy and D'Cruz, 1966), and hybridization induced, facultative, aposporous embryo sac production has been recorded in *Raphanobrassica* (Ellerstrom and Zagorcheva, 1977). The role of hybridization and/or polyploidy in inducing asexuality was first proposed by Ostenfeld (1912) for plants, and subsequently by numerous authors, including White (1973) and Suomalainen and Saura (1973), for animals. Muntzing (1940) proposed what Cuellar (1987) describes as the 'genetic imbalance' hypothesis to account for the relationship between asexuality and these factors. Here, apomixis is considered to be due to a delicate genetic balance that may be upset by hybridization or by changes in ploidy level.

However, even though there is some evidence that hybridization and/or polyploidy can generate asexual phenotypes, there is little evidence that asexuality typically results from these factors alone. Indeed, both polyploidy and hybridization are common among obligately sexual taxa, indicating that they do not provide a ready route to asexuality for many organisms. These and similar observations have led numerous authors to propose that, although hybridization and polyploidy may assist in the production of asexual phenotypes, they are usually not sufficient in themselves to generate such phenotypes. The expression of apomixis genes is required for this (e.g. Gustafsson, 1936; 1946–47). (Apomixis genes are defined as genes which control aspects of the apomictic phenotype – parthenogenesis, the avoidance of meiotic reduction etc. – and which are typically absent from, or at very low frequencies in, sexual populations.) Nevertheless, the controversy over whether hybridization and/or polyploidy have been responsible for most instances of asexuality is one that has been difficult to put to rest. As Cuellar (1987) wryly observes, the debate is no sooner satisfactorily resolved with regard to one major taxon than it emerges anew in another. Thus, with respect to the role of hybridization in generating asexual phenotypes, he states (Cuellar, 1987: 68)

> ... the controversy germinated in plants in 1912 (Ostenfeld, 1912) was carefully tendered in the same group in 1917 and 1918 (Winge, 1917; Ernst, 1918), was very much alive in moths in 1926 (Peacock and Harrison, 1926), accelerated its pace in beetles in the seventies (White, 1970; Suomalainen and Saura, 1973), and is thriving today in grasshoppers (Hewitt, 1975; White *et al.*, 1977).

From about the 1940s onwards, the view of many researchers has been that the switch from sexual to asexual reproduction will typically require the acquisition of specific 'asexual' (or 'apomixis') genes. One of the first genetic models was offered by Powers (1945), who proposed a three-locus control for generative apomixis. One locus controlled the avoidance of meiotic reduction, one the avoidance of fertilization, and one the parthenogenetic development of the embryo. The genes controlling these processes were considered to be recessive to those controlling sexuality. This model has attracted considerable criticism, mostly because of its proposal that an apomixis gene is required for the avoidance of fertilization (e.g. Gustafsson, 1946–47). Although Powers' model was offered as a general one, its development was influenced by his investigations into the control of apomixis in *Parthenium argentatum*. Thus, it is interesting to note that Gerstel *et al.*

(1953) have proposed an alternative scheme for this species, with apomixis being controlled by at least four recessive genes. A minimum of two are concerned with the avoidance of meiotic reduction, and two more with the control of parthenogenesis.

Models involving two or more loci or sets of loci are common. Asker (1980) argues that such loci are independent and are brought together in the sexual ancestors of apomicts by mutation or recombination. Grant (1981) provides basically the same argument, suggesting that the apomixis-determining alleles of the separate genes or gene systems could be carried as unexpressed recessive factors in the ancestral sexual population. Usually, one locus is assumed to control parthenogenesis. The role of the other varies with the pattern of apomixis. For example, in generative apomicts it may control the avoidance of meiotic reduction, whereas in aposporous apomixis it may be responsible for the development of the aposporous embryo sac.

With respect to apospory, a two-locus control has been suggested for *Cenchrus ciliaris* (*Pennisetum ciliare*) by Taliaferro and Bashaw (1966), who argue that apospory is controlled by an epistatic interaction between these loci, one of which codes for sexual reproduction and the other for apospory. The presence of a gene for sexual reproduction results in a sexual phenotype, irrespective of the allelic composition of the locus which codes for apospory. Thus apospory results from the loss through mutation of the genes coding for sexual reproduction. Hanna *et al.* (1973) have proposed that a minimum of two loci are involved in the control of apospory in *Panicum maximum*. They argue that sexuality is dominant to apomixis and that it is conditioned by a dosage effect of two or more dominant alleles at two or more loci. Savidan (1975, 1980, 1981, 1982), however, suggests that the control of apospory is probably not as complicated as this in this species, and that it comprises a single supergene that controls both aposporous embryo sac production and parthenogenesis and that apospory is dominant to sexuality. Harlan *et al.* (1964) have argued that apospory in *Dichanthium* is controlled by a dominant gene. D'Cruz and Reddy (1971) consider the control to be more complex in this genus, with genes for apomixis and for sexuality being non-allelic and not dominant to one another. Burton and Forbes (1960) propose that apospory in *Paspalum notatum* is recessive to sexuality and is controlled by a small number of genes. Nogler (1984) proposes that aposporous embryo sac formation in *Ranunculus auricomus* is controlled by a single mutant gene which shows some dominance but is susceptible to dosage effects of the wild type gene that codes for sexual reproduction. The control of parthenogenesis in this species is unknown, but is unlikely to be due to simple Mendelian genes (Nogler, 1984).

Richards (1973, 1986) has proposed a two locus (or sets of loci) control for generative apomixis in *Taraxacum*, with the genes for apomixis being dominant to those for sexual reproduction. One controls the avoidance of meiotic reduction, the other controls parthenogenesis and precocious embryony. Rutishauser (1946) suggests that generative apomixis in *Potentilla* is controlled by a single recessive gene which can achieve phenotypic dominance through dosage effects. Two copies of this gene dominate one copy of the wild-type (which codes for sexual reproduction). Rutishauser (1947) further proposes that apospory in this genus is also caused by a recessive but different gene.

It is clear from this brief and incomplete review that there is little agreement on the nature of the control of apomixis. This may, in part, be due to different taxa exhibiting different mechanisms of control. But in several cases (see above) different models have been proposed for the same taxon. This confused situation has led Nogler (1984: 499) to comment that 'A definitive statement about the causes of gametophytic apomixis cannot yet be made, as many more details need to be collected.'

However, it is my belief that sufficient data are available to allow the construction of a robust model of how both generative and aposporous forms of apomixis are controlled. The major premise on which this model will be based is that the various components that must be brought together to generate an apomictic phenotype are to be found occurring separately in a wide range of sexually reproducing organisms. Often, more is known about these processes from investigations of sexual organisms than from investigations of apomicts. By bringing data together from both groups, sufficient can be amassed to indicate how apomixis is controlled.

5.3 THE COMPONENTS OF APOMIXIS AND THE NATURE OF THEIR CONTROL

A major problem with evolving apomictic reproduction is that its components are individually highly deleterious. An organism that acquires the capacity for parthenogenesis but which continues to produce eggs by reductional meioses will produce weak or inviable haploid, or even possibly subhaploid, progeny. Similarly, an organism that acquires the capacity to produce meiotically unreduced eggs but which lacks the capacity to develop these parthenogenetically will initiate a lineage that will rapidly polyploid itself out of existence as eggs are fertilized. Only when the individual components are brought together in a phenotypically expressible form will they contribute to a selectively advantageous asexual phenotype. The problem here is that individually highly deleterious genes are only likely to persist in a population if they are recessive, and even then only at low frequency (Falconer, 1981). Thus if the evolution of a selectively advantageous character such as apomixis requires the simultaneous bringing together of several of these mutations in phenotypically expressible form it is unlikely to evolve.

This problem suggests *a priori* that the control of apomixis in at least the majority of apomicts is likely to be simple, requiring at most the bringing together of a very small number of mutations. Thus an initial step in the development of the model is to consider which, if any, of the components of apomixis are likely to require the acquisition of a mutant gene (apomixis gene) that is rare in, or absent from, sexual populations. However, the model must do more than this if it is to be generally applicable. It must also be able to explain the variation observed in the way different organisms achieve apomictic reproduction. Some of the variation in the way meiotic reduction is avoided has already been described in Chapters 2 and 3. However, there are other variable characteristics. Thus the amount of synapsis initiated during prophase I of the restitutional meiosis undertaken by apomicts undergoing the *Taraxacum* scheme varies. Asynapsis is typical, but synapsis has been observed in some taxa, including *Parthenium* (Rollins, 1945), *Rubus nitidoides* (Haskell, 1953), *Agropyron scabrum* (Hair, 1956) and *Taraxacum palustre* (Malecka, 1965). Also, in flowering plants, variation is seen in the number of mitotic divisions involved in aposporous embryo sac formation, and in whether apomixis is facultative or obligate, or aposporous or generative. With respect to the last point, the model must establish whether aposporous and generative apomixis are, at the level of control, fundamentally different or similar processes. It must also explain how both the capacity for sexual reproduction and that for asexual reproduction are retained by facultative apomicts. Moreover, it must explain how the relationship between these capacities varies so that one achieves phenotypic dominance in some ovules and the other in the remainder. Finally, it is important to obtain some insight into the extent to which the observed variation in the pattern of apomixis is due to genetic differences between organisms at loci involved in the control of apomixis, rather than to other causes.

There is one other factor that the model must consider. This is the relationship between apomixis and polyploidy. Almost all of the aposporous plants and generatively apomictic plants and animals that have been cytologically examined are polyploid. This is the case even when all related non-apomictic forms are diploid (Nygren, 1967; Suomalainen *et al.*, 1976; Bell, 1982; Nogler, 1984). This relationship between polyploidy and apomixis extends to facultatively apomictic forms as well as to obligately apomictic forms. Several explanations for this association have been given. It has been argued that polyploidy is advantageous because of its effects on the ecological range and tolerance of apomicts (Stebbins, 1950), because it provides surplus loci which can buffer an apomict against the effects of deleterious mutations while allowing it to generate advantageous mutations (Ohno, 1970; Lokki, 1976a,b) or because it increases the mutation rate per individual towards an optimum (Manning and Dickson, 1986). However, although these effects of polyploidy may be advantageous, there are two reasons for doubting that they provide the primary reason for its prevalence among generative and aposporous apomicts (Mogie, 1986b). The first is that the proposed advantages of polyploidy are ones that reduce the costs of uniparental reproduction in general. Consequently, if generative and aposporous apomicts have, almost without exception, been selected to acquire polyploidy then so should organisms exhibiting other forms of uniparental reproduction. And yet this is not the case. Though common, polyploidy is not prevalent among flowering plants reproducing by adventitious embryony or among parthenogenetic, automictic animals. It could be argued that taxa that can evolve generative and aposporous apomixis can acquire polyploidy more easily than taxa evolving other forms of uniparental reproduction, and that this explains the discrepancies in its distribution. This is a reasonable argument but its relevance is fatally weakened by Bell's (1982) observation that the incidence of polyploidy can vary between closely related taxa exhibiting different forms of uniparental reproduction. This is the case in the Tardigrada (microscopic aquatic animals allied to arthropods) and *Artemia* (shrimps). Both contain automictic as well as generatively apomictic species. Within the Tardigrada, automicts may be diploid or polyploid but apomicts are invariably polyploid; within *Artemia*, all automicts are diploid, whereas all apomicts are polyploid. Such observations provide a clear indication that polyploidy is closely associated with generative and aposporous apomixis rather than with uniparental reproduction *per se*. The second reason for doubting that the ecological and mutational benefits generated by polyploidy are the primary reasons for its prevalence among apomicts is that if this was the case then polyploidy should not be as prevalent among facultative apomicts as it is among obligate apomicts, as the former have recourse to biparental sexual reproduction to increase their ecological and evolutionary potential. And yet it is as common. For example, whereas all obligately sexual *Taraxacum* species are diploid, all apomictic species, whether obligately or facultatively so, are polyploid (Richards, 1973). These anomalies are large enough to cast doubt on current explanations for the association between polyploidy and generative and aposporous apomixis. One alternative explanation is that polyploidy aids the expression of the apomictic phenotype. This has been proposed by Rutishauser (1946) for *Potentilla* and it is an explanation that will be extended to generative and aposporous apomicts in general in section 5.4.

The problem of whether apospory and generative apomixis are, at the level of control, fundamentally different or similar processes has intrigued researchers for many years. For example, Nogler (1984: 507) states that 'It is an open question whether the different forms of diplospory are, from a genetic point of view, fundamentally different from apospory, or whether there is a common basis.' This is a view shared by Bergman (1951) and Urbanska-Worytkiewicz (1974). Certainly, the usually distinct taxonomic demarcation between apospory and generative apomixis breaks down on occasion, indicating that they may simply be different manifestations of the same basic process. For example, although most *Antennaria* apomicts are generatively apomictic, apospory as well as generative apomixis occurs in *A. carpatica* (Bergman, 1951). Other species which exhibit both include *Potentilla verna*, *Artemisia nitida*, *Parthenium argentatum*, *P. incanum* and *Rubus saxatilis* (Esau, 1946; Nygren, 1954; Czapik, 1981). Similarly, several genera contain both aposporous and generatively apomictic species, including *Poa*, *Elatostoma*, *Potentilla* and *Paspalum* (Nygren, 1954; Smith, 1963; Chao, 1974, 1980; Nogler, 1984).

The problem concerning the extent to which variation in the pattern of apomixis reflects differences between organisms at loci involved in the control of apomixis is one that has received little attention. Nevertheless, there are several strands of evidence that indicate that the apomictic phenotype can show considerable environmentally induced plasticity against a uniform genetic background. This evidence relates to three areas: variation in the pattern of generative apomixis, variation in the pattern of apomixis *per se*, and variation in the proclivity of facultative apomicts to reproduce sexually.

The first reported case of an environmentally induced switch in the pattern of generative apomixis was supplied by Ernst and Bernard (1912), who observed a switch from the usual *Antennaria* type of embryo sac production to the *Taraxacum* type in *Burmannia coelestis*. Sparvoli (1960) describes a similar switch in *Eupatorium riparium*. Nogler (1984) suggests that such changes may be frequent.

Several examples have already been provided of taxa which exhibit both aposporous and generative forms of apomixis. The extent to which these different apomictic phenotypes reflect different mechanisms of control or simply environmental effects is unclear. However, a well-documented case of how environmental effects can alter the apomictic phenotype, even to the extent of replacing one type of apomixis with another, is provided by Reddy and D'Cruz (1969), who investigated this matter in facultatively aposporous *Dichanthium annulatum*. As there is no reason to suppose that this taxon is unique in this respect, it is pertinent to describe this investigation in some detail.

Reddy and D'Cruz (1969) investigated intra-individual variation in embryo sac structure and in the pattern of avoidance of fertilization in *D. annulatum*, which does not exhibit precocious embryony and, as we shall see, does not exhibit precocious oogenesis, at least in many ovules. In this taxon, the megaspore mother cell frequently degenerates during meiosis. It was noticed that, either accompanying or following this degeneration, one or more nucellar cells in the ovule enlarged and developed into aposporous embryo sacs. However, the aposporous sacs could attain either a four-nucleate or an eight-nucleate state, with both types frequently being observed within the same ovule. In addition to aposporous embryo sacs, adventitious embryos were observed in some ovules. These formed directly from nucellar cells and frequently invaded the aposporous embryo sacs (presumably making use of the sac's endosperm). Thus two different patterns of aposporous embryo sac formation and both apospory and adventitious embryony were recorded within and between ovules of a single individual – that is, within the confines of a single genotype!

Considerable intra-individual variation was also demonstrated between aposporous embryo sacs in the events surrounding embryo initiation. *D. annulatum* is a pseudogamous apomict, with the development of an adequate nutritive endosperm tissue relying on the fertilization of a polar nucleus. Because each pollen grain deposits two male gametes into an embryo sac, this requirement for fertilization of the polar nucleus carries with it the risk that the egg may be fertilized by the surplus male gamete. Four pathways were identified by which this risk is minimized. Once again, different pathways were followed by embryo sacs occupying the same ovary, that is by embryo sacs exhibiting the same genotype. In four-nucleate aposporous embryo sacs, the single polar nucleus is invariably fertilized by one of the male gametes. The remaining male nucleus is free to enter the egg and frequently does so. It does not invariably do so, however, and the exclusion of the male gamete from the egg comprises the first pathway by which these apospores avoid egg fertilization. Eight-nucleate embryo sacs do not necessarily face the problem of having to deal with a 'spare' male gamete. These sacs contain two polar nuclei. In 14 out of 27 such sacs observed after pollination, each male gamete fused with a polar nucleus. The double fertilization of the polar nuclei comprises the second pathway by which these apospores avoid egg fertilization. However, in 13 of the 27 eight-nucleate sacs only one male gamete ferti-

lized a polar nucleus. The other was potentially free to enter the egg, and in nine cases it did so. Thus, irrespective of whether the aposporous embryo sac is four or eight nucleate, there is a high probability that a male gamete will enter the egg. However, in these instances, the male nucleus fails to fuse with the egg nucleus. Instead, it degenerates immediately, or after one or a few divisions. This failure of syngamy comprises the third pathway by which the fertilization of the egg is avoided. The fourth pathway, which was observed in only a single ovule, is one which allows fertilization to be effectively, though not actually, avoided. This ovule contained a six-celled aposporously derived embryo. An examination of the number of nucleoli in these cells indicated that one of the two cleavage nuclei of the embryo had been fertilized. This cell divided to give rise to the suspensor, which although ontogenetically part of the embryo, degenerates during embryo maturation. The unfertilized cleavage nucleus gave rise to the embryo proper. (It should be pointed out that although the authors found no evidence of aposporous eggs being fertilized in this experiment, in a further experiment (D'Cruz and Reddy, 1971) they used a tetraploid apospore as a female parent and obtained pentaploid progeny. This suggests that a meiotically unreduced egg was fertilized by a meiotically reduced male gamete and that the avoidance of fertilization is therefore not absolute in aposporous embryo sacs of this species.)

This example clearly illustrates that whatever the nature of the control of apospory, its pattern can vary within the confines of a single genotype. Reddy and D'Cruz (1969) did not identify the environmental factor responsible for the variation. However, it is presumably generated in response to variation in the physiological and biochemical environments both of different parts of an ovule and of different ovules. Certainly, this type of variation has been implicated in determining the extent to which facultative apomicts, including *D. annulatum*, reproduce sexually – the aspect of variation which will be considered next. Given that it can have this extreme effect, it is probable that it will also cause the types of modifications to the apomictic process documented in this species.

Environmentally induced variation in the proportion of sexually produced embryos has been recorded for several facultative apomicts. Before referring directly to these, it is apt to consider the changes that must be undertaken to generate this variation. Basically, there is a requirement for fundamental changes in the pattern of ovule development. In facultative forms reproducing asexually by generative apomixis, variation in the proportion of sexually and apomictically produced progeny largely reflects variation in the proportion of megaspore mother cells that undergo a reductional meiosis, as sexually produced

embryos result from the fertilization of reduced eggs and apomictically produced progeny result from the parthenogenetic development of unreduced eggs. In facultative forms reproducing asexually by aposporous apomixis, variation in the proportion of sexually and apomictically produced progeny largely reflects variation in the survivorship of megaspore mother cells or their immediate products, as aposporous embryo sacs in many taxa appear to develop only in ovules in which the megaspore mother cell or its immediate products degenerate.

With respect to both forms of apomixis, variation in the frequency of sexually to apomictically produced progeny has been described above as 'largely' reflecting changes in ovule development. This note of caution is sounded because an alternative explanation for the observed changes in frequency of the two types could be differential survival rates, rather than differential initiation rates, of apomictic and sexual embryo sacs. This appears to be the explanation for an increase in the proportion of sexually produced progeny in the generative apomict *Eragrostis curvula* in which, under short photoperiod days, meiotically unreduced megaspores have a tendency to degenerate whereas meiotically reduced ones initiate viable embryo sacs (Brix, 1974 and personal communication in Nogler, 1984). Nevertheless, in a number of apomicts, changes in the frequency of the two types do seem to reflect changes in ovule development. This appears to be the case in the generative apomicts *Calamagrostis purpurea* (Nygren, 1946) and *Eupatorium riparium* (Sparvoli, 1960) and in the aposporous apomicts *Dichanthium annulatum* (Knox and Heslop-Harrison, 1963; Gupta *et al.*, 1969), *D. intermedium* (Saran and de Wet, 1970, 1976), *D. aristatum* (Knox, 1967) and *Themeda triandra* (Evans and Knox, 1969). In *Dichanthium* and *Themeda*, short photoperiods favour apospory over sexual reproduction. For example, *D. aristatum* produced aposporous embryo sacs in 91% of its ovules when the photoperiod was less than 14 hours, but in only 60% of its ovules when the photoperiod was greater than this (Knox, 1967). In *D. intermedium*, apospory could be completely suppressed by altering the length of the photoperiod (Saran and de Wet, 1976).

It is clear that a great deal of variation in the apomictic phenotype need not imply a great deal of variation in the genetic control of apomixis – it has been shown above how variation can be generated within the confines of a single genotype by altering the environment. There is another potential source of variation. This is the genetic background against which genes for apomixis must operate. If, across individuals, apomixis genes could be kept constant but genetic background was allowed to vary, the changes in genetic background would affect the ovule environment and would consequently also probably affect the apomictic phenotype. Thus reports of intra- or inter-specific

differences in the pattern of apomixis should not necessarily be interpreted as evidence that different apomicts exhibit fundamentally different mechanisms of control of this reproductive process.

The preceding discussion has demonstrated that not all variation in the apomictic phenotype is due to differences in genotype. Attention will now be shifted towards consideration of which components of apomixis are likely to demonstrate an absolute requirement for apomixis genes. The components which are most relevant to this discussion are the avoidance of meiotic reduction, the production of aposporous embryo sacs, the avoidance of fertilization, and the production of embryos by parthenogenesis. The avoidance of fertilization through precocious oogenesis/embryony has already been dealt with in Chapter 3, where it is proposed that precocity is simply a consequence of the change in timing of egg maturation that results from the abbreviated pathway of megaspore production exhibited by apomicts. That is, precocity can be attained without the acquisition of a specific mutation for precocity. The other components will now be considered. Parthenogenesis will be addressed first.

Before embarking on this, it is pertinent to emphasize now that what I am attempting to find is the most parsimonious route for the evolution of apomixis. This is an easily defended aim. A recurring theme in this book is that asexual reproduction is only likely to be found in those taxa which can acquire it easily. It is shown in Chapter 3 how bryophytes and gymnosperms cannot acquire it easily and how it is consequently absent from these groups. Asexuality needs to be acquired easily because it cannot be acquired piecemeal: parthenogenesis and the production of meiotically unreduced eggs are characteristics that are advantageous when they are exhibited together but they are deleterious when one is expressed in the absence of the other. The opinion has already been expressed in this chapter that if their acquisition requires the accumulation of separate mutations then apomixis is unlikely to evolve. But if this line of argument is to have any validity it is essential that evidence can be marshalled in its support. The evidence may indicate that only some taxa can acquire one or more of these components without mutation. This is not a problem. It simply means that apomixis is more likely to evolve in these than in taxa which can only acquire them by mutation. This matter will be returned to in the next section.

The control of parthenogenesis

Eukaryotes can be divided into three categories with respect to parthenogenesis: (1) those which cannot initiate embryos parthenogenetically; (2) those which can initiate embryos but are unable to mature them; and (3) those which can produce mature, viable embryos by parthenogenesis. Homosporous plants fall firmly within the first category, mammals within the second. The third category includes many insects and flowering plants.

The inability of homosporous plants to initiate embryos parthenogenetically is a consequence of the nature of the chromatin of the egg; this is so finely dispersed that, in the absence of a nuclear template, it is unable to condense sufficiently to allow the egg to divide mitotically. The nuclear template is provided by the more condensed chromatin of the male gamete, thus condemning homospores to non-parthenogenetic forms of reproduction (Bell, 1979a, and section 3.3).

The mammalian egg is not constrained in this way and expresses a potential for initiating embryos parthenogenetically (Graham, 1974). However, the resulting embryos appear to be incapable of developing to maturity. It is becoming clear that for a mammalian embryo to reach maturity it must contain genes that have been donated by a sperm as well as maternal genes (Surani *et al.*, 1984, 1986, 1988; McGrath and Solter, 1984; Solter, 1987; Jackson, 1989). It seems that parts of the genome are genetically imprinted by passage through egg or sperm and that this imprinting affects their activity in the embryo. The pattern of imprinting is different in eggs and sperm so that both a maternal and a paternal genome are required for normal embryo development. Thus in mice, diploid embryos with nuclei constructed from two haploid female genomes or from two haploid male genomes are inviable, whereas those with one of each are viable. This demonstrates that inviability is not simply a consequence of haploidy and that viability is not simply a consequence of diploidy. Rather, diploidy is associated with viability because it is usually a consequence of the bringing together of maternal and paternal genes. Possibly, differential imprinting of maternally and paternally donated genes results from differences in DNA methylation patterns in eggs and sperm (Reik *et al.*, 1987; Swain *et al.*, 1987).

It is clear that in homosporous plants and in mammals, the evolution of a viable system of parthenogenetic reproduction will require complex changes to the way in which DNA is organized and modified during egg production. There is no reason for homosporous plants to accumulate the mutations necessary for these changes as there is a cost of asexual reproduction associated with homospory (Chapter 3). However, there is a potential twofold fitness advantage to parthenogenetic reproduction among mammals. Consequently, its absence from this group indicates that no mammal has yet succeeded in either acquiring the mutations necessary for the transition from sexual to asexual reproduction, or in acquiring them in the right sequence in a phenotypically expressible form. This takes the discussion back to the view that asexual reproduction is only likely to evolve if its acquisition requires the accumulation of only a very small number of mutations. It is therefore of great interest to note that in the two major taxa in which parthenogenetic reproduction is a common method of asexual reproduction – flowering plants and insects – there are no obvious barriers preventing an unfertilized egg from dividing. Indeed, there is a great deal of evidence that parthenogenesis is an innate capacity in many insects, being induced by phenomena that are intrinsic to the reproductive process. For example, Went (1982) postulates that egg deformation during oviposition can induce parthenogenesis in members of the Hymenoptera (bees, wasps, ants), which is one of the largest orders of the animal kingdom, comprising some 100 000 species, most of which are haplodiploid. Events associated with oviposition may also trigger parthenogenesis in the grasshopper *Melanoplus differentialis* (King and Slifer, 1934) and in *Drosophila melanogaster* and *D. mercatorum* (Went, 1982). Exposure to atmospheric oxygen after oviposition is the trigger for egg cell division in the apomictic stick insect *Carausius morosus* (Pijnacker and Ferwerda, 1976).

The role of oviposition in stimulating egg cell division in the Hymenoptera has been investigated by Went (1975, 1982; Went and Krause, 1973, 1974) in a series of experiments involving the haplodiploid endoparasitic Ichneumonid *Pimpla turionellae*. In this species, ripe but unfertilized eggs are retained in the ovarioles until a few seconds before oviposition and injection of the eggs into the host pupa. During these few seconds the eggs are first discharged into the oviduct, where a proportion are fertilized, and then the egg mass is injected into the host. Fertilized eggs develop into diploid females, unfertilized eggs into haploid males. Eggs were obtained both from ovarioles and after oviposition to determine when egg activation occurs. Those removed from the ovarioles did not develop, whereas oviposited eggs did. Obviously, fertilization cannot be implicated as the stimulus for development, as only a proportion of the oviposited eggs that developed had been fertilized. The diameter of the egg canal in the ovipositor is only about one-third the width of the eggs; the ovipositor is constructed of solid material and is not likely to yield to a passing, pliable egg. It was concluded that eggs must get greatly distorted during their passage down through the ovipositor and that this might activate them. In order to test this, eggs were removed from ovarioles and, in an elegantly simple yet powerful experiment, were distorted by passing them through capillaries, mimicking oviposition. Over 70% of the eggs so treated were activated.

Went (1982) argues that eggs may be commonly activated by distortion in many members of the Hymenoptera. In some species, eggs emerge through the vagina without having been distorted by the ovipositor (which has lost its egg-laying function and has been converted into a sting). In these species egg distortion may occur in either the uterus or the vagina.

Went's observations and experiments lend considerable support to the view that parthenogenesis in many taxa is an innate capacity that can be expressed in the absence of 'apomixis genes'. Fertilization may activate eggs, but so too will numerous other phenomena, such as distortion during ovipositioning. Flowering plants lack an obvious mechanism for distorting their eggs. However, such physical manipulation is only one of a number of stimuli that are known to induce parthenogenesis. The problem is not so much one of attempting to identify parthenogenetic stimuli but is one of coming to a decision about whether the angiosperm egg is commonly receptive to such stimuli. The knowledge that many sexual taxa are able to initiate haploid embryos, either spontaneously or following artificial stimulation (Kimber and Riley, 1963) provides strong evidence that this is the case. This opinion receives further support from the arguments presented in Chapter 3 concerning the evolution of the embryo sac. Evidence was presented that the angiosperm egg is homologous to an ancestral archegonial cell (the ventral canal cell) that retained mitotic potential, rather than to the homosporous egg which, we have seen, lacks this potential. There is no obvious reason why, following this substitutional process, there should have been selection for the angiosperm egg to lose this potential, which will be expressed as an innate capacity for parthenogenesis. Taking these factors into account, it appears reasonable to assume, in the absence of firm evidence to the contrary, that parthenogenesis in this group is a commonly held capacity. Moreover, the association between precocious embryony and parthenogenesis which was highlighted in Chapter 3 indicates that this capacity will be expressed in many taxa if fertilization is avoided or delayed.

In conclusion, there is little reason to accept, and considerable reason to reject, the hypothesis that parthenogenesis in insects and flowering plants will function as a component of apomixis only following the acquisition of an 'apomixis' gene. Rather, the evidence leans heavily in favour of parthenogenesis being an innate capacity in at least many members of these groups, and consequently in the ancestors of at least many apomicts.

The control of egg production in apomicts

The discussion concerning parthenogenesis has gravitated towards the view that a viable system of parthenogenesis is commonly within the normal range of abilities of insects and flowering plants. The evidence is circumstantial and the discussion has been left somewhat up in the air. But it will not be left there. It will be grounded within the framework of the ensuing discussion, the immediate aim of which is to consider the nature of the control of the meiotic process in apomicts. This issue is relevant to aposporous as well as to generative apomixis. However, in order to avoid unnecessary complication, I will deal with these consecutively rather than concurrently. Attention will first be focused on generative apomixis.

Most generative apomicts avoid meiotic reduction either by substituting a mitosis for meiosis (*Antennaria* scheme) or by replacing the synaptic reductional meiosis with a restitutional one that is wholly or largely asynaptic (*Taraxacum* scheme) (details are given in Chapter 2). There is a considerable body of evidence that these changes are mediated by apomixis genes. The firmest evidence comes from a comparison of the course of male and female meiosis in cosexual apomicts. Whereas female meiosis is replaced by the *Antennaria* or *Taraxacum* schemes, male meiosis is typically synaptic and reductional, though often highly disturbed. These disturbances include univalent and multivalent formation. They can lead to irregular disjunction and result in unreduced and partially reduced, as well as fully reduced, male gametes. They are to be expected. Apomictic flowering plants are polyploid, often anisoploid and (on morphological and cytological evidence) often hybrids. It is reasonable to suspect that these factors cause the disturbances observed during male meiosis. Certainly, they are of a type often observed in raw hybrids or polyploids of sexual species. The significance of this is that the much more regular course of the restitutional or mitotized female meiosis cannot be dismissed as a consequence of hybridization or polyploidy, or indeed of any other non-specific event that could upset the course of a reductional, synaptic meiosis (e.g. chromosome mutation). It must result from the action of genes – apomixis genes – that affect female, but not male, meiosis.

No such genes have been positively identified in generative apomicts. However, a family of genes that disrupts meiosis and that can generate both *Taraxacum* type restitutional and *Antennaria* type mitotized meioses has been identified in a taxonomically diverse array of sexual organisms, including insects and plants where generative apomixis is a common form of asexual reproduction. These genes – meiotic mutations – have been located at a large number of loci. Detailed descriptions of them are provided by Baker *et al.* (1976) and Lewin (1980). Their effects vary considerably. Many cause a partial disruption of a reductional meiosis, mostly through reducing, though not eliminating, synapsis. However, a few are of particular interest to this discussion. For example, *ameiotic* in maize switches the division from a meiotic to an *Antennaria* type mitotic course, and *as* in maize causes asyndesis. Some, including *as3* in *Brassica campestris*, generate a *Taraxacum* type division, causing both asyndesis and a meiosis which comprises only a single division, and a number influence female but not male meiosis, including *elongate* in maize and a number of types in *Drosophila*. Of considerable interest is the observation that the phenotypic effects of these mutations can be modified by genetic background. For example, the *as* mutation in maize generates phenotypes which range from completely syndetic in some plants to almost completely asyndetic in others. Most importantly, an organism need possess only a single mutation for its female meiosis to become asyndetic and restitutional or replaced by a mitosis.

It is clear that these meiotic mutations present themselves as strong candidates for apomixis genes. They provide a range of phenotypes that encompass those exhibited by generative apomicts. Moreover, the phenotype a mutation generates can be subject to modification by the genetic background. Most importantly, many meiotic mutations have been identified at many loci and yet only a single locus needs to be affected for meiosis to be switched onto an apomictic course, and a number of these mutations affect female but not male meiosis. Thus, although the difference in the pattern of male and female meiosis in cosexual apomicts provides evidence that the avoidance of meiotic reduction during egg production is controlled by apomixis genes, investigations of meiotic mutations in sexual taxa provide evidence that only one such gene, arising at any one of a number of loci, may be needed to generate this facet of the apomictic phenotype. As evidence has already been presented that apomixis genes may be unnecessary for the avoidance of fertilization and for parthenogenesis in insects and flowering plants, the conclusion to be drawn is that generative apomixis in these taxa can result from the acquisition of a single mutation. The details of how this can be achieved will be discussed in section 5.4. Before this, attention will be switched to aposporous apomixis. Consideration of this process demonstrates that meiotic mutations may play a role in its control similar to that played in the control of generative apomixis.

In apospores, the embryo sac is derived mitotically from a somatic cell of the ovule rather than from the megaspore mother cell (Chapter 2). Thus for apospory to comprise an efficient method of reproduction, four requirements must be met: an aposporous embryo sac must be produced; this must replace the meiotically derived embryo sac; the egg of the aposporous embryo sac must initiate an embryo parthenogenetically; this embryo must be viable. The last two of these requirements have already been discussed earlier in this chapter, where it was concluded that neither necessarily requires the acquisition of apomixis genes. Attention will consequently be focused on the first two requirements.

In most apospores, the initiation of the aposporous embryo sac coincides with the degeneration of the megaspore or of the embryo sac derived mitotically from this (Nogler, 1984). It is still not clear whether degeneration results from innate inviability or is a consequence of competition with the aposporous embryo sac. Nogler considers that the former explanation is correct for the majority of apospores, citing as evidence observations from Warmke (1954) on *Panicum maximum*, Brown and Emery (1957) on *Themeda triandra* and his own observations on *Ranunculus auricomus* (Nogler, 1971) that megaspores degenerate in ovules in which aposporous embryo sacs have failed to develop. Although this conclusion is based on observations of only three species, there is other evidence that points to its validity. Moreover, this evidence indicates why it is typically the megaspore, rather than an earlier or later stage, that degenerates. This evidence concerns some members of the family of meiotic mutations implicated earlier in the control of generative apomixis, and their effects on the course of egg and early embryo development in *Drosophila*.

Lewin (1980) describes three different meiotic mutations that cause the degeneration of the egg or early embryo in *Drosophila*. The mutant *pal* causes the loss of paternal chromosomes at meiosis and at the first mitoses of the embryo. The mutant $ca^{(nd)}$ causes non-disjunction and chromosome loss at oogenesis; it can also cause the loss of maternal chromosomes during the first embryonic mitosis. The mutant *mit* causes the loss of both maternal and paternal chromosomes at the third or fourth mitotic divisions of the embryo. Lewin suggests that the wild-type genes at these loci code for some components of the chromosomes that are necessary for the first few post-meiotic mitoses and that are synthesized during meiosis. After the first few divisions these components are synthesized by the mitotic cells themselves. These components may involve the centromeres or the spindle apparatus (Davis, 1969; Baker and Hall, 1976; Lewin, 1980). Interestingly, these mutations can also cause non-disjunction at the first meiotic division (Baker *et al.*, 1976).

The presence of similar mutations in flowering plants would have different effects that would make them very attractive candidates for apospory genes. Namely, they would cause the degeneration of the megaspore or its embryo sac, as the first few post-meiotic mitoses in flowering plants are involved in megaspore division leading to embryo sac formation rather than in embryo initiation. Their predicted action is therefore that seen in apospores. Indeed, the tight linkage between their effects and meiosis provides an answer to a nagging problem, namely why an aposporous embryo sac is able to follow a viable course of development when the megaspore-derived embryo sac in the same ovule is unable to do so. The answer is simply because the mitoses involved in the production of the megaspore-derived embryo sac are post-meiotic ones, whereas those involved in the production of aposporous embryo sacs are not – they are 'normal' (i.e. not post-meiotic) mitoses. As it is a failure during meiosis to produce a substance required for a successful completion of the first few post-meiotic mitoses, degeneration is only to be expected in megaspore mother cell-derived megaspores or embryo sacs.

The fact that a single family of mutations can be identified that generates phenotypes associated with both generative and aposporous apomixis is significant as it provides the genetic link between these two processes that has for long been sought and predicted. Of considerable significance is the observation that the same mutation may cause the types of disruption to the meiotic process associated with both aposporous and generative apomixis, offering an explanation for the phenomenon described earlier of both processes being associated with some taxa.

Although the meiotic mutations described appear to be prime candidates for apomixis genes there are problems associated with assigning them this role. These are surmountable and will be discussed in the next section. However, there is one final issue associated with the control of apospory that first needs to be considered. This concerns the control of the initiation of the aposporous embryo sac. As a rule, this is initiated from a (somatic) nucellar cell that adjoins the chalazal pole of the megaspore, and its initiation in most taxa is associated with the degeneration of either the megaspore or the embryo sac derived from it. This observation allows the tentative prediction to be made that the aposporous embryo sac is initiated in response to substances that control embryo sac development but which are produced by the megaspore and released during or following its degeneration. These substances will accumulate at high concentration around the degenerating megaspore/embryo sac and will therefore be more likely to trigger aposporous embryo sac development in adjoining, rather than in distant, somatic cells. Thus the initiation of an aposporous embryo sac may simply be in response to the degeneration of the megaspore or its embryo sac rather than in response to the expression of an apomixis gene different from that causing degeneration. Consequently, apospory, as well as generative apomixis, can result from the acquisition of a single apomixis gene that affects the course of the female meiosis. This explanation cannot be supported with experimental evidence. But nor can the less parsimonious alterna-

tive explanation that a separate apomixis gene is required for the
initiation of the aposporous embryo sac.

5.4   A MODEL FOR THE CONTROL OF GENERATIVE AND
APOSPOROUS APOMIXIS

Setting the background

It is clear from the foregoing discussion that a viable apomictic phenotype could be generated by a meiotic mutation. But acquiring one of
these mutations is clearly not sufficient *per se* to guarantee an apomictic phenotype. Otherwise, we would not know of their existence from
sexual organisms! Thus detailed consideration must be given to
whether or not apomixis has usually resulted solely from the acquisition of one of these mutations.

On acquiring a meiotic mutation a sexual organism will become
apomictic only if it exhibits innate parthenogenetic potential and only
if it can produce an egg in a way that allows this innate potential to
be expressed. But although an innate capacity for parthenogenesis may
be widespread there is no evidence to suggest that most organisms will
be able to express it following the acquisition of a meiotic mutation. It
is argued in the previous section that meiotic mutations could have
this effect, through pleiotropically inducing precocious egg development or aposporous embryo sac production. But it cannot be concluded
from this that they will usually have this effect. Indeed, it is possible
that only a few taxa will be stimulated by the degeneration of a
megaspore or its products to initiate embryo sacs aposporously, or
that only a few taxa will be able to produce eggs from meiotically
unreduced megaspores with sufficient precocity to allow viable parthenogenetic embryos to be produced. Taxa with these attributes
could be described as being pre-adapted to evolving aposporous or
generative apomixis easily. But taxa which lack these attributes will
have to acquire mutations for aposporous embryo sac development or
for a viable pattern of precocious embryony in addition to acquiring
a meiotic mutation if they are to evolve apomixis. Unfortunately,
there is no direct observational evidence that can be used to provide
an estimate of the proportion of sexual taxa that are pre-adapted for
apomixis. In the absence of this it would be wrong to assume that
pre-adaptation is common.

Although there are insufficient data available to quantify the problem there are data and arguments that can be used to illustrate it. For
example, Stebbins (1941) describes individuals of several species which
can produce embryo sacs aposporously but which appear to be unable
to initiate embryos parthenogenetically. These include individuals

of *Oxyria digyna*, *Antennaria dioica*, *Coreopsis bicolor* and *Picris hieracioides*. Similarly, in *Chrysanthemum carinatum* the megaspore mother cell may degenerate during meiosis (Bergman, 1952) but this does not stimulate the development of aposporous embryo sacs: degeneration leads to sterility, not to apomixis. In these examples pre-adaptation for apomixis appears to be absent: *Chrysanthemum carinatum* will need at least a mutation for aposporous embryo sac production in addition to a meiotic mutation if it is to reproduce as an aposporous apomict; the other species will need a mutation for parthenogenesis if they are to reproduce as aposporous apomicts.

Even when apomictic tendencies are present they may not be exhibited in a manner conducive to the expression of a viable apomictic phenotype. This can be appreciated by reconsidering an aspect of generative apomixis: precocious oogenesis/embryony. It is shown in Chapter 3 that the most parsimonious explanation of this phenomenon is that it is a pleiotropic consequence of the mutations controlling the avoidance of meiotic reduction during egg formation. It occurs because these mutations accelerate the process of egg maturation to such an extent that the egg is able to express an innate capacity for parthenogenesis before male gametes are able to gain access to it. However, because precocity results from pleiotropy, the imbalance it generates in a newly arisen meiotic mutant will rarely be optimal. The imbalance may be insufficient to prevent fertilization, or it may be so great that the parthenogenetically initiated embryo develops before there is an endosperm adequate to support it. This latter problem will be accentuated in those taxa in which endosperm development is dependent on the polar nuclei being fertilized, that is, in those taxa which are potential pseudogamous apomicts. It is argued in Chapter 3 that problems in achieving the right level of precocity during egg production may account for the uneven taxonomic distribution of *Taraxacum*-scheme apomicts and *Antennaria*-scheme apomicts within the flowering plants, as these two schemes generate different levels of precocity. The *Taraxacum* scheme is the most taxonomically restricted of the two, being largely confined to the family Asteraceae. The pattern of megaspore production it is associated with also engenders the least precocity. Presumably members of other families acquiring this level of precocity would find it insufficient to ensure that most eggs avoided fertilization. That is, members of the Asteraceae are more likely than members of other families to be pre-adapted to evolving apomixis on the acquisition of a meiotic mutation that causes a restitutional *Taraxacum*-type division. Members of other families are likely to require mutations for precocious oogenesis/embryony in addition to a mutation for a *Taraxacum*-type division if they are to evolve a pattern of apomixis involving this division, although members of some families (e.g. the Poaceae) may be pre-adapted to evolving apomixis on the acquisition of a meiotic mutation that induces an *Antennaria*-type division.

Even when a taxon may be pre-adapted in terms of precocious oogenesis/embryony for the evolution of generative apomixis the acquisition of an optimal pattern of apomixis may be difficult. Most newly initiated lineages may exhibit a viable but inefficient pattern of apomixis. This can be seen most clearly in terms of the regulation of pseudogamous apomixis. The level of pleiotropically induced precocity in these must be sufficient to preclude egg fertilization but not so great that the endosperm develops too late to support the embryo. But even when the level of precocity meets this requirement it may be suboptimal. Thus if the embryo is initiated before the optimum period the endosperm may be able to supply it with sufficient resources to maintain its viability but may be too immature to supply these in quantities that will optimize the embryo's growth rate. Conversely, if the embryo is initiated after the optimum period, the opportunity for rapid or early growth may be missed as the embryo may be too immature to take full advantage of the resources the endosperm is capable of providing it with. In each case, the lineage may persist and improve the efficiency of apomictic reproduction by accumulating modifier genes that move the timing of embryo initiation towards the optimum. However, it is important to stress that these modifier genes will be selected after apomixis has evolved. They are selected to improve this pattern of reproduction, not to initiate it! A distinction between initiating and modifier genes must therefore be made when investigating the genetic control of apomixis.

It is easy to conclude from this type of deliberation that whereas apomixis could result solely from an organism acquiring a meiotic mutation it is unlikely to do so. In most taxa, more than one mutation will be required if apomixis is to evolve. Given this, it may appear that I'm embarking on a fool's errand in the task which I am about to begin, which is to formulate a model for a single gene control of generative and aposporous apomixis and then test its validity against the available data. But I undertake this task in the firm belief that it is a far from pointless exercise. I can most easily defend this view by referring back to some of the arguments that were discussed earlier in this book.

In Chapter 3, it is argued that the absence of asexuality from the bryophytes and gymnosperms is due to asexual reproduction being too difficult to evolve. Its establishment in the bryophytes will require the acquisition of a gene for female sterility as well as the acquisition of a gene for asexual reproduction, and there may be additional requirements for the acquisition of genes that will result in the asexual mutants achieving ecological isolation from sexual conspecifics. Its establishment in the gymnosperms will require the acquisition of genes that prevent pollination as well as the acquisition of a gene for asexual reproduction. Even so, it is unlikely that the absence of asexuality in these two groups is due to them being unable to generate the necessary mutations. It is much more likely that its absence is due to these mutations being individually deleterious and to the difficulty of bringing them together in one individual at the same time and in phenotypically expressible form. It is as valid to apply this basic argument to the flowering plants as it is to apply it to the bryophytes and the gymnosperms: each of the separate components of apomixis (meiotic restitution, parthenogenesis etc.) in flowering plants is individually deleterious and they must be brought together at the same time and in phenotypically expressible form if a potentially advantageous apomictic phenotype is to be expressed. The probability of this happening is extremely small if each of the components must be acquired separately by mutation. In contrast, the probability of apomixis evolving will be much higher if its evolution requires the acquisition of only a single mutation. It has already been shown how this evolutionary pathway, involving the acquisition of a meiotic mutation, is a feasible one. The taxa which can utilize this pathway may comprise only a small minority of flowering plant taxa but they will be over-represented among apomicts – at least if one condition applies, namely that the time scale over which apomixis could have evolved is of sufficient length to have favoured its acquisition only by those taxa which are pre-adapted to evolving it by acquiring only a single mutation. In general, if acquiring a selectively advantageous character by mutation takes, on average, x years in some taxa but $4x$ years in others, then if these taxa were to be examined $2x$ years after their initiation we would find the character exhibited by many of the first set of taxa but by few of the second set. The first set may represent a minority of the taxa, but they will make up the bulk of those taxa that exhibit the character. But if we were to return after $8x$ years we would find that the character was ubiquitous within both sets of taxa.

Some indication of where in the time scale the observation is being made can be obtained by examining the distribution of the character. If the capacity for acquiring the character is influenced by taxonomic location the character will be initially restricted to the taxa most pre-adapted to acquiring it; let us say to those that need to obtain only a single mutation to acquire it. But as we move further along the time scale we will reach the point at which taxa that need two mutations to acquire the character will acquire it, and so on. Thus, with time, the character will become both more common and more evenly distributed among taxa. This point is important, as the taxonomic distribution of apomixis *per se*, and of the different forms of apomixis, is very restricted within the flowering plants and its frequency, both overall and in the families in which it is found, is low. Thus it is shown in Chapter 2 that parthenogenetic apomixis is found in only 7% of families and in less than 1% of species, and Richards (1986) has calculated that 75% of such apomicts belong to only three families – the Asteraceae, Poaceae and Rosaceae – which together comprise only about 10% of angiosperm species. Ecological factors and life history characteristics will explain some of this variation, as apomixis may be more selectively advantageous in some situations than in others. Thus it is much more common among perennial herbs than it is among either annual herbs or perennial woody taxa, and it is more common in temperate areas and in grasslands than in other areas. It will therefore be more frequent in those families that are common in temperate and grassland areas and that contain a high proportion of perennial herbs. However, such factors explain only part of the variation observed. Another cause of variation will arise through some higher taxa having evolved broad ontogenetic patterns that are conducive to the evolution of apomixis. It appears that most taxa are not pre-adapted in this way and, from its distribution, it appears that these have not been in existence for long enough for the two or more mutations they need to evolve apomixis to have been acquired simultaneously by an individual in phenotypically expressible form. That is, extant taxa of flowering plants have been in existence only for long enough for apomixis to have been acquired by those taxa which are pre-adapted to acquire it easily. In my view, these will comprise those taxa in which its evolution can result solely from the acquisition of a meiotic mutation.

The uneven taxonomic distribution of apomixis is one of two factors that suggest that extant taxa which are not pre-adapted to evolve apomixis easily have not been in existence for long enough to acquire it by the more difficult routes available to them. The second factor concerns the irregular distribution of generative apomixis and apospory within those families in which parthenogenetic apomixis is found. For example, generative apomixis is common among apomicts of the Asteraceae but is less so among apomicts of the Poaceae, whereas apospory is common among apomicts of the Rosaceae and of the Poaceae but is less so among apomicts of the Asteraceae. Similar differences persist at lower taxonomic levels. For example, within the Asteraceae apospory is the most common form of apomixis in *Crepis* and in *Hieracium* subgenus *Pilosella*, but generative apomixis is the most common form in *Hieracium* subgenus *Hieracium* and is the only form in *Taraxacum* (Richards, 1986). Similarly, within the Poaceae apospory is the typical form of parthenogenetic apomixis within the

tribe Andropogoneae, with the exception of *Saccharum* which is generatively apomictic, whereas generative apomixis is typical of apomicts
of the tribe Agrostideae, although all apomictic members of this tribe
belong to one genus – *Calamagrostis* (Connor, 1979). There is no *a
priori* reason to assume that these differences reflect differences
between taxa in the rates of acquisition of mutations generating the
meiotic disturbances associated with each type of apomixis. It is more
likely that they reflect differences between taxa in rates of development or in the susceptibility of somatic ovule cells to stimulation,
resulting in different taxa being able to utilize the different types of
mutation differently. A taxon may be pre-adapted with respect to
precocious oogenesis/embryony and parthenogenesis to utilizing meiotically unreduced megaspores in generative apomixis, but it does not
follow from this that it will also be pre-adapted to produce aposporous
embryo sacs if megaspores or their immediate products were to
degenerate. This latter pre-adaptation will require the somatic cells of
the ovule to be stimulated to divide mitotically by the products of
degeneration. In the absence of this response additional mutations
will be required to generate aposporous embryo sacs. Thus, in such
taxa, a single mutation will be required for the evolution of generative
apomixis, but two or more will be required for the evolution of apospory. The proportion of generative apomicts should consequently be
much greater than that of aposporous apomicts.

The model

It is not the purpose of this model to describe the steps that must be
taken by a random sample of sexual individuals if these are to achieve
apomixis. Rather, its purpose is to offer an explanation of how apomixis has evolved in generative and aposporous apomicts. It is very
important that this distinction is maintained: as we have just seen, it
is probable that taxa that have evolved efficient patterns of apomictic
reproduction are derived from a non-random subset of sexual taxa that
were pre-adapted to evolve apomixis easily. The steps involved in the
evolution of apomixis in these will be fewer than those that would
need to be involved in taxa that are not pre-adapted for its evolution;
the latter group may comprise the majority of sexual taxa.

The first stage in the development of the model involves an examination of the consequences of assigning meiotic mutations the role of
apomixis genes. Although they are obvious, indeed the only, candidates for this role there are difficulties in assigning it to them. In
overcoming these much can be learned about several aspects of the
control of apomixis. The problems in assigning meiotic mutations the
role of apomixis genes become apparent if the mode of action of these

genes in sexual taxa and the expression of apomixis in apomictic taxa are examined.

In sexual taxa, meiotic mutations are recessive to the wild type genes coding for normal syndetic and reductional meioses and/or normal postmeiotic mitoses (Baker *et al.*, 1976). Consequently, it would be expected that an apomict would need to be homozygous for a meiotic mutation. But this would result in it being obligately asexual. And yet facultative apomixis is a common form of apomixis among flowering plants. Facultative apomicts have retained the capacity for normal syndetic and reductional meioses and consequently must have retained at least one copy of a wild-type gene at the locus affected by the meiotic mutation. Thus there is a problem in explaining how the recessive mutation finds phenotypic expression, if the locus at which it resides also contains a copy of the dominant wild-type gene.

A second problem concerns deleterious secondary effects that have been associated with a number of meiotic mutations. Briefly, a number of these mutations affect mitosis in somatic cells as well as meiosis in female generative cells. Their effects on mitosis can take several forms including a disruption of the pattern of growth and development, an increase in the rate of point and chromosome mutation and/or an increase in the rate of mitotic recombination (Baker *et al.*, 1976; Lewin, 1980). Although they are absent, there is no immediately obvious reason why these deleterious effects should not be expressed in the somatic cells of apomicts possessing meiotic mutations.

Acknowledging these two problems forces the conclusion that if meiotic mutations are to function as apomixis genes they must be held with wild-type genes at heterozygous loci but in such a way that they attain phenotypic dominance over the wild types under a restricted set of conditions. These conditions are that they achieve phenotypic dominance in some or all ovules, thereby imparting a pattern of reproduction that is facultatively or obligately apomictic, but that they do not achieve phenotypic dominance in somatic cells. Only if the last condition is fulfilled will the organism be able to follow a viable pattern of growth and development.

These conditions can be met if two reasonable assumptions are made. The first is that the dominance relationship between a mutant gene and its wild type can be reversed by dosage effects. This will allow the mutant phenotype to be expressed in heterozygotes which contain excess copies of the mutant gene. Note that the easiest way to achieve this dosage effect is through polyploidy, offering an explanation for the strong positive association between generative and aposporous apomixis and this condition. Other mechanisms include amplification of the affected locus, providing multiple copies of the locus in an otherwise diploid genome. This will be more difficult to evolve than polyploidy and will therefore be less common. But it may explain why a few apomictic taxa are diploid.

The second assumption is that the environment of somatic cells and, in facultative apomicts, of some ovules favours the dominance of the wild type gene over its mutant forms, even if the latter are present in excess numbers. That this assumption is reasonable can be seen by referring back to the previous section where it was described how in facultative apomicts the phenotypic dominance of the tendencies towards apomixis and sexual reproduction is environmentally labile, and is so on a scale that is sensitive enough to result in dominance reversals distinguishing ovules of an individual. Within the boundaries of the present discussion, this can be interpreted as demonstrating that the assumed dosage effect can be nullified by a change in the cellular environment. Of course, whereas the phenomenon of facultative apomixis demonstrates that the dominance relationship between the wild-type and mutant alleles can be labile in ovules, the validity of the model is dependent on the wild-type alleles permanently exhibiting dominance in somatic cells. This requires that the environment of somatic cells is very different from that of ovules. But there is little doubt that this is the case. In preparation for meiosis, the megaspore mother cells of flowering plants experience changes in their internal and external environment that are both profound and unique to them. The changes to this cell include a disruption of nutrient supply and the disappearance of most of its organelles (Kapil and Bhatnagar, 1981). Bouman (1984) makes the point that this cell becomes effectively, though temporarily, isolated from the rest of the plant just before and during meiosis, enabling it to embark on an independent course of development. This isolation, which results from the deposition of callose on the cell wall, occurs at the time the meiotic mutations will be having their disruptive effects on the course of meiosis.

Given these two assumptions, it can be seen that an individual that is homozygous for a mutation will be inviable because of the disruptive effects of this gene on growth and development; a heterozygote with one copy of the mutant gene and one of the wild type will be phenotypically sexual, because of the recessive nature of the mutant gene at this dosage level; but a heterozygote with excess copies of the mutant gene will be a viable apomict. The genotypes and predicted phenotypes of diploid and triploid carriers of a meiotic mutation are described in Table 5.1 (tetraploid carriers will be investigated later). It can be seen from this that of the seven possible genotypes only one (genotype 6, *Aaa*, where *A* is the wild-type gene and *a* the mutant gene) generates a fully viable apomictic phenotype.

Table 5.1 The genotypes and predicted phenotypes of diploid and triploid carriers of a meiotic mutation. Only one genotype (genotype 6) produces a fully viable apomictic phenotype

No.	Genotype	Phenotype
Diploid carriers of meiotic mutation		
1	AA	Sexual
2	Au	Sexual
3	aa	Apomictic but weak or inviable through the disruption of growth and development
Triploid carriers of meiotic mutation		
4	AAA	Sexual
5	AAa	Sexual
6	Aaa	Apomictic
7	aaa	Apomictic but weak or inviable through the disruption of growth and development

A = wild type; a = mutant allele

In summary, it has been shown in this and previous sections that the control of generative and aposporous apomixis can reside at a single locus, the identity of which can vary between lineages. These loci contribute to the control of meiotic reduction. The wild-type genes at these loci help to mediate a normal, syndetic, reductional meiosis and normal postmeiotic mitoses. The mutations at these loci can have several effects, including imparting complete or partial asyndesis on the meiotic division, switching it onto a mitotic course, and disrupting the first few postmeiotic mitoses. This accounts for the variation observed between apomicts in the pattern of apomixis. The model proposed for the control of apomixis states that the affected locus contains at least one copy of the wild-type gene and excess copies of the mutant gene. It is argued that the dominance relationship between these is determined by their ratio and by the environment, with environmental differences between ovules in some individuals being sufficient to cause reversals of dominance and therefore cause facultative apomixis. It is further argued that environmental differences between female generative cells and somatic cells are such that the phenotypic expression of the mutant allele is favoured in the former whereas the phenotypic expression of the wild-type is favoured in the latter. This is important, for the locus is also involved in the control of mitosis which would be disrupted by the expression of the mutant allele in somatic cells. The requirement to maintain a viable pattern of growth and development explains why the locus is heterozygous. The requirement for excess copies of the mutant allele in generative cells explains why generative and aposporous apomicts are typically polyploid, as this condition provides a simple and effective means of generating the correct balance of mutant and wild-type alleles. The control of apomixis needs only to reside at a single locus because the other components of apomixis do not require the acquisition of apomixis genes in pre-adapted taxa. Parthenogenesis is an innate capacity of many organisms; the avoidance of fertilization through precocious oogenesis/embryony is a consequence of the earlier maturation of the egg that follows the abbreviated course of megaspore/egg initiation which is itself a consequence of the expression of the meiotic mutation; and the initiation of aposporous embryo sacs can be stimulated by the degeneration of the megaspore or its products, an event which is induced by the meiotic mutation.

5.5   VERIFICATION AND FINE-TUNING OF THE MODEL

The model proposed has been developed mostly from consideration of apomictic tendencies (parthenogenesis, meiotic restitution etc.) in non-apomictic organisms. However, for it to be at all valid it must not contradict what little we know from apomicts about the control of this process. Most of this knowledge has been obtained from manipulative experiments. The data generated by these can be interpreted in several ways and, as described earlier in this chapter, have frequently been interpreted as demonstrating a multi-locus control of apomixis. However, in no case have these interpretations been verified by detailed genetic analysis. In this section, several of these data sets will be reanalysed to see if they fit the more parsimonious single-locus model. They have been obtained from investigations of the control of both aposporous and generative apomixis in triploids and tetraploids. No distinction will be made in the reanalysis between the two types of apomixis, as the model encompasses both. But ploidy level will have a major determining effect on the nature of the reanalysis and will be duly respected. The more straightforward reanalysis of data obtained from triploid apomicts will be offered first. Throughout, the assumptions that have determined the directions and affected the conclusions of previous analyses will be examined, and the assumptions on which the reanalyses are based will be explained. It will be shown that even a simple genetic model can lay barriers in the path of the analyst attempting to locate the correct null hypothesis and its alternatives. The aim of this section is not simply to test the validity of the single locus model. It has two further purposes. One is to demonstrate that apomicts are often in a position to accumulate apomixis genes that are surplus to requirements. The other is to attach the genetic model to a possible biochemical framework.

Apomixis in triploid *Taraxacum*

Most *Taraxacum* apomicts are triploid, although there are some tetraploid taxa. Apomixis is non-pseudogamous and generative, and most apomicts are obligately so although there are some facultatively apomictic forms. Obligately sexual forms are diploid. Thus there is a clear demarcation between ploidy level and the pattern of reproduction. The significance of this demarcation has been illustrated by Richards (1970a,b; 1973) who investigated the inheritance of apomictic behaviour among progeny derived from crosses between diploid sexuals (used as females) and triploid apomicts. The eggs produced by the diploid sexuals will have been haploid. The pollen produced by the apomicts showed various levels of meiotic reduction, from fully reduced to fully unreduced. As a result, the progeny exhibited a range of ploidy levels. Those which were diploid or near diploid were sexual. Some which were triploid or near-triploid were also sexual, but others were apomictic. Thus it appears that progeny from this type of cross will be apomictic only if they receive more than a single haploid genome from the apomictic parent, although the receipt of more than one haploid genome does not guarantee apomixis.

These results can be explained in terms of the model if it is assumed that the triploid parents exhibited genotype *Aaa*, which Table 5.1 shows to be the only fully viable apomictic genotype for triploids. It is immaterial whether the sexual diploids are considered to be homozygous for the wild-type gene or heterozygous. The range of genotypes and phenotypes resulting from crosses of this type are illustrated in Table 5.2. No diploid genotype will produce a viable apomictic phenotype, and only one triploid genotype (genotype 4, *Aaa*), which is identical to that of the apomictic male parent, and one tetraploid genotype (genotype 10, *Aaaa*), will definitely do so. Whether or not the tetraploid genotype *AAaa* (genotype 5) will produce an apomictic phenotype is open to debate, and this issue will be discussed below. At least one polyploid genotype, *AAa*, produces a sexual phenotype.

Further information on the nature of the control of apomixis in this genus is provided by Malecka (1965), Sørensen and Gudjónsson (1946) and Sørensen (1958). Malecka demonstrated that sexuality can re-emerge in an obligately apomictic lineage following the gain of chromosomes by non-disjunction. The apomict investigated was eutriploid ($2n = 3x = 24$) but occasionally generated progeny with one or more extra chromosomes, through a process of non-disjunction. Most of these were obligately apomictic like their mother. However, female meiosis in one, with two additional chromosomes, had reverted from the asynaptic, restitutional pattern found in the mother to one that was synaptic and fully reductional. Malecka determined that the two

Table 5.2 The genotypes and phenotypes resulting from a cross between a diploid sexual with genotype *AA* or *Aa* (acting as female parent) and a triploid apomict with genotype *Aaa*. Male meiosis in the apomict is reductional but highly disturbed, resulting in pollen that is fully reduced (*A* or *a*), partially reduced (*Aa* and *aa*) and unreduced. Meiosis in the sexual is reductional and undisturbed, resulting in *A* eggs and *a* eggs

	Egg	Pollen	Progeny genotype	Progeny phenotype
1	*A*	*A*	*AA*	Sexual
2	*A*	*a*	*Aa*	Sexual
3	*A*	*Aa*	*AAa*	Sexual
4	*A*	*aa*	*Aaa*	Apomictic
5	*A*	*Aaa*	*AAaa*	Could be apomictic or sexual depending on the nature of the dosage effect (see text)
6	*a*	*A*	*Aa*	Sexual
7	*a*	*a*	*aa*	Inviable due to the disruptive effects of the mutant gene on growth and development
8	*a*	*Aa*	*Aaa*	Apomictic
9	*a*	*aa*	*aaa*	Inviable due to the disruptive effects of the mutant gene on growth and development
10	*a*	*Aaa*	*Aaaa*	Apomictic

extra chromosomes were homologous. It is probable that they represented sister chromatids of the same chromosome, in which case they would carry the same alleles at corresponding loci. Unfortunately, this plant died before seed set but all of the indications were that it had fully reverted to sexuality. The results of this investigation are important because they demonstrate that the capacity for sexuality appears to be retained, though phenotypically suppressed, in obligate apomicts of this genus, and that the putative re-emergence of this capacity resulted not simply from an increase in chromosome number *per se* but from an increase in the number of copies of a particular chromosome. This indicates that the re-emergence of sexuality is due to a dosage effect involving one or more loci on this chromosome rather than being simply a consequence of aneuploidy. In terms of the model, if it is assumed that the parent had the genotype *Aaa* and that the two extra homologous chromosomes acquired through non-disjunction each carried a copy of the *A* allele, then the genotype of this individual would be *AAAaa*. This would shift the balance of alleles at this locus in favour of the *A* allele, resulting in it being phenotypically expressed and in meiosis reverting to the syndetic reductional form observed.

Sørensen and Gudjónsson (1946; Sørensen, 1958) investigated another type of chromosome mutation that caused obligately apomictic lineages to express a capacity for sexual reproduction. This

mutation involved the loss of a chromosome, so that obligately apomictic eutriploid $(2n = 24)$ parents gave rise parthenogenetically to
aberrant $2n = 23$ hypoploid progeny. Such progeny were obtained
from several taxa. Most of the hypoploids obtained were obligately
apomictic like their parents but a few were partially sexual, producing
some progeny sexually and some parthenogenetically. Once again, this
indicates that the switch in phenotypes was not simply due to a
change in chromosome number *per se* but to the loss of a particular
chromosome or chromosomes, providing further evidence for a genetic
control of apomixis.

Sørensen and Gudjónsson attempted to identify the chromosome
which, when lost, caused sexuality to re-emerge. They concluded that
it could be one of either of two non-homologous morphological types.
They therefore divided the partially sexual aberrants into types *elegans*
and *tenuis* depending on which chromosome type they had lost.
Crosses were made in which the hypoploid sexual aberrants were
pollinated by compatible sexual and apomictic taxa. A few of the
progeny resulting from these crosses were cytologically analysed. It
was found that those that had been derived sexually had arisen following the fertilization of both meiotically reduced and meiotically unreduced eggs and that those that had been derived parthenogenetically
had arisen from meiotically unreduced eggs. This pattern of embryo
initiation differs from that found in non-aberrant facultatively apomictic members of this genus; in these, only meiotically reduced eggs are
fertilized, meiotically unreduced eggs developing parthenogenetically.
Thus the loss of a chromosome in these hypoploids did not simply
result in a switch from obligate apomixis to a typical form of facultative apomixis. The data from this cytological analysis are summarized
in Table 5.3.

Female meiosis in *Taraxacum* apomicts is restitutional and largely
or wholly asyndetic, unlike male meiosis which is frequently synaptic
and reductional though often highly disturbed. It seems, from the
chromosome numbers of their progeny (Table 5.3), that at least some
of the female meioses in the partially sexual aberrants had switched
to a reductional, synaptic, male-type course. This was confirmed for
one aberrant. Indeed, it is likely that all the female meioses in these
aberrants were essentially reductional but that, because of disturbances, some produced unreduced or only partially reduced products.
Evidence in support of this statement emerges from consideration of
column 4 of Table 5.3, which shows that some of the progeny derived
parthenogenetically from meiotically unreduced eggs (the maternal
types) contained up to three more chromosomes than their mothers.
Further evidence is provided by Sørensen's (1958) observation that
parthenogenetically derived progeny were more variable morphologi-

Table 5.3 A summary of Sørensen's (1958) cytological data on the partially sexual *Taraxacum* aberrants. The female parents are partially sexual aberrants. The male parents are either sexual or apomictic. 'Maternal type' progeny are considered by Sørensen to be derived parthenogenetically from meiotically unreduced eggs. Hybrid progeny are derived sexually by the fertilization of meiotically reduced and meiotically unreduced eggs. Note that the pollen of the sexual parent will be $n = 8$, but that of the apomictic parents will vary from approximately $n = 8$ to approximately $n = 24$, because of the disturbed nature of the reductional male meiosis typical of many apomicts of this genus

Female parent	Male parent	Plants analysed	Maternal types (2n = 23–26)	Chromosome numbers of hybrid progeny (2n)							
				16	17	18	23 to 25	29 to 31	30	31	39
cordatum aberr. *tenuis*, 2n = 23	*obtusilobum* (sexual) 2n = 16	9	2	–	3	2	–	–	1	1	–
cordatum aberr. *tenuis*, 2n = 23	*polyodon* aberr. *tenuis* 2n = 23	4	1	1	1	1	–	–	–	–	–
polyodon aberr. *elegans*, 2n = 23	*bracteatum* (apomict) 2n = 24	8	4	–	–	–	4	–	–	–	⋯
polyodon aberr. *elegans*, 2n = 23	*hamatum* (apomict) 2n = 24	7	2	–	–	–	2	3	–	–	–
polyodon aberr. *elegans*, 2n = 23	*chloroleucum* (apomict) 2n = 24	5	2	–	–	–	3	–	–	–	–
laciniosifrons aberr. *elegans*, 2n = 23	*cyanolepis* (apomict) 2n = 24	7	0	–	–	–	–	4	–	–	3

cally than expected (i.e. were more variable than the progeny of normal apomicts). Both phenomena could result from a synaptic but restitutional meiosis, with enhanced morphological variation resulting from recombination, and with extra chromosomes resulting from nondisjunction either during the second meiotic division or during the mitoses involved in embryo sac or early embryo development.

One interesting aspect of this investigation is that the production of meiotically reduced eggs by the aberrants resulted in the re-emergence of sexual reproduction rather than in the emergence of haploid parthenogenesis. The fertilization of these eggs demonstrates that precocious embryony (found in apomicts of the genus), the parthenogenetic production of viable embryos, and the avoidance of meiotic reduction are somehow intertwined. However, the fact that some meiotically unreduced eggs were fertilized indicates that this association is not absolute. This is to be expected within the framework of the arguments presented in Chapter 3 and earlier in the present chapter. It is argued in Chapter 3 that precocious embryony in apomicts is achieved as a consequence of the pattern of megaspore production, rather than as a consequence of the acquisition of a mutation that acts specifically to achieve precocity. It is argued earlier in this chapter that parthenogenesis in apomicts is an innate capacity which will find expression if fertilization can be avoided through mechanisms such as precocity. Thus precocity and parthenogenesis are largely linked, not to meiotic restitution *per se*, but to restitution when it results from an abbreviated pathway of megaspore production. This abbreviated pathway appears to have been lost in the sexual hypoploids, with female meiosis reverting to the male-type that incorporates syndesis and therefore a non-abbreviated prophase I. The fact that some meiotically unreduced eggs developed parthenogenetically, even though they appear to have been produced by a non-abbreviated meiosis, may reflect accelerated development resulting from their unreduced status – a phenomenon also discussed in Chapter 3. Basically, a meiotically unreduced egg may develop faster than a reduced egg. However, the rate of development of unreduced eggs is not sufficiently accelerated to ensure that they avoid fertilization. To ensure this, these eggs must also be initiated earlier than meiotically reduced eggs. This requires an abbreviated or mitotized meiosis. In *Taraxacum*, precocity works because it causes the egg to mature before the flower opens to allow the stigma to be pollinated. It appears that in this genus unreduced eggs produced by a non-abbreviated meiosis mature around the time the flower opens. Some will mature before pollination has been effected and will develop parthenogenetically. Others will mature slightly later and will be fertilized.

To return to the main theme of this discussion, Sørensen and Gudjónsson considered that the sexual types *elegans* and *tenuis* were morphologically distinct. This provides some supporting evidence that the two types had lost different loci and thus had lost non-homologous chromosomes. However, the sample size on which the authors based this conclusion is totally inadequate. One of the main species used in their investigation – *T. polyodon* – produced only one *elegans* and one *tenuis* aberrant (these aberrants were used as females in the crosses described in Table 5.3). In all, only five *tenuis* aberrants seem to have been obtained from three species. Two of these species also produced a total of seven *elegans* aberrants (Sørensen and Gudjónsson, 1946). This sample size of 12 aberrants from three species is inadequate for any conclusions to be drawn on morphological differences between *elegans* and *tenuis* aberrants. Sørensen and Gudjónsson also stated that the two aberrants differed in their behaviour as females, with *elegans* type reproducing mostly by the parthenogenetic development of meiotically unreduced eggs, and with *tenuis* types reproducing mostly by the fertilization of meiotically reduced eggs. As each type is reported as lacking a copy of a different chromosome type, Sørensen and Gudjónsson's conclusions on both cytology and reproductive behaviour appear to contradict the model proposed in the previous section for the control of apomixis. Basically, because both *elegans* and *tenuis* aberrants produced meiotically reduced as well as unreduced eggs, Sørensen and Gudjónsson's conclusions indicate that the asynaptic and restitutional meiosis typical of apomicts of this genus is dependent on interactions between loci on these two chromosomes. In the absence of these interactions (i.e. following the loss of one or other of these chromosomes) meiosis reverts to being synaptic and, barring disturbances, reductional. Moreover, either the frequency of restitution or the capacity for parthenogenesis also appears to be influenced by loci on both chromosome types, as differences in the frequency of either could account for the reported differences in reproductive behaviour between the two types. Sørensen and Gudjónsson did not attempt to produce a model for the control of apomixis in this genus. However, Richards (1973) concluded that their data indicate that the missing chromosome in the *elegans* aberrants carries a dominant gene for parthenogenesis that is expressed only in cells at the triploid level or above, and that the missing chromosome in the *tenuis* aberrants carries a dominant gene that causes meiotic reduction to be avoided. However, this model can be challenged as both *elegans* and *tenuis* aberrants produced meiotically reduced as well as meiotically unreduced eggs, and both types also produced some progeny parthenogenetically. This is clear from the chromosome numbers listed in Table 5.3. It is also clear from this Table that Sørensen and Gudjónsson's conclusions on the reproductive behaviour of the two types are based on small sample sizes and should therefore be viewed with caution. Only 13 of the progeny obtained after crossing one *tenuis* aberrant were cytologically examined, as were only 27 of the progeny obtained after crossing *elegans* aberrants.

There is a further reason to doubt Richards' (1973) model, and indeed to doubt the cytological accuracy of Sørensen and Gudjónsson's (1946) investigation on which it is based. Basically, the cytological analysis performed by Sørensen and Gudjónsson is suspect (Richards, 1986; Mogie, 1988). They began their investigations by analysing the karyotype of a diploid (2*n* = 16) sexual species, which they determined contained two copies of each of eight morphologically distinguishable chromosomes. They then appear to have assumed that all *Taraxacum* species contain these eight chromosome types and that the triploid apomicts they were examining were autoploid and therefore contained three copies of each type. As a result, eutriploid apomicts were all given identical karyotypes, and their hypoploids were then 'assigned' a missing chromosome and grouped accordingly. However, subsequent examinations have shown that apomicts of this genus are alloploid (Gustafsson, 1946–47), that the genus contains more than the eight types of chromosome identified by Sørensen and Gudjónsson, that individual triploid apomicts do not contain three morphologically identical haploid genomes (Malecka, 1962, 1965, 1967a,b, 1969, 1970, 1971, 1972; Richards, 1972; Singh et al., 1974; Mogie, 1982), and that one of their apomictic species (*T. hamatum*) contains two morphologically distinct satellite chromosomes rather than the three Sørensen and Gudjónsson (1946) identified (Mogie and Richards, 1983). It is also known that at least some of the chromosomes of at least some *Taraxacum* apomicts, including *T. hamatum*, are prone to structural rearrangement (Mogie and Richards, 1983; Richards, 1989). Thus it is best to discount that part of Sørensen and Gudjónsson's (1946) observations that emanate from their classification of chromosome morphology, and thus to discount their division of the partially sexual hypoploid aberrants into types *elegans* and *tenuis*.

In doing so, the only clear message emerging from Sørensen and Gudjónsson's investigations is that sexuality can re-emerge in an obligately apomictic lineage following the loss of a chromosome. This is in agreement with the model developed in the previous section. Consider that the obligately apomictic parents had genotype *Aaa* (the only viable apomictic genotype according to the scheme presented in Table 5.1). The loss of a chromosome carrying one or other of these alleles would produce genotypes *Aa-*, *A-a* and *-aa*. The first two of these would generate sexual phenotypes (which would, however, be disturbed because of the hypotriploid chromosome number), the third an inviable phenotype. Thus, once the doubtful parts of Sørensen and

Gudjónsson's analysis have been rejected or re-interpreted, their data can be shown to fit the single locus model proposed here for the control of apomixis.

Tetraploidy gene dosage and apomixis

In the previous section and subsection, it is argued that a simple excess of apomixis genes at a locus is sufficient to generate an apomictic phenotype in triploids. Simply extending this argument to the tetraploid case will force the conclusion that tetraploid apomicts will exhibit the genotype *Aaaa* (using the previous notation) at the relevant locus. However, the situation may not be as simple as this. Under certain patterns of chromosome pairing and gene product construction, it can be shown that an *AAaa* genotype, but not the diploid *Aa* genotype which exhibits the two genes in the same ratio, will generate an apomictic phenotype. This suggests that in tetraploids the total number of copies of mutant and wild-type alleles as well as their ratio may determine whether a sexual or an apomictic phenotype is produced. Supporting data are provided by *Dichanthium*. Before these are introduced, the problem will be examined in general terms, by taking a closer look at the nature of the mutations involved in the disruption of meiosis or of the postmeiotic mitoses.

In sexual organisms, meiotic mutations that cause asyndesis and restitution prevent or impair the formation of the synaptonemal complex; those that cause the disruption of postmeiotic mitoses appear to affect some component of the centromeres or of the spindle apparatus (Baker *et al.*, 1976; Lewin, 1980). It is probable that the mutations that lead to the *Taraxacum* scheme of apomixis and to apospory have these types of effect. The detailed biochemical effects of these mutations are unknown but it can be speculated that they interfere with some aspect of the construction of structural proteins or with some aspect of the structure and activity of enzymic proteins, as both types of protein are involved in the construction of synaptonemal complexes and of meiotic and mitotic spindles.

What follows is a biochemical interpretation of the control of apomixis. It is based on the observation (e.g. Fincham, 1983) that many proteins comprise two or more functional polypeptide components (monomers). It is highly speculative, but if its sole attribute is to stimulate debate then it will serve a useful purpose. It has the advantage of offering both a biochemical and a quantitative explanation for the delicate balance observed in some apomictic flowering plants between apomixis and sexuality.

The basic biochemical unit that will be considered is a dimer – a protein comprising two monomers. Assume initially that the locus controlling apomixis controls the synthesis of this dimer. In order to avoid the proliferation of descriptive terms, the monomer produced by the wild-type *A* allele will be described as the *A*-monomer, and that produced by the mutant *a* allele will be described as the *a*-monomer. Both monomers are produced by a heterozygote. That is, although the *A* and *a* alleles may exhibit dominance and recessivity with respect to each other at the level of the organism (i.e. the phenotype is either *A* or *a*) they are co-dominant at the level of protein synthesis. The wild-type *A*-monomer is functional (active), the mutant *a*-monomer inctive. A heterozygote will produce three types of dimer: *AA*, *Aa* and *aa*. It will be assumed that the *AA* and *Aa* dimers are active, the latter because of the presence of the *A*-monomer. It will further be assumed that more active than inactive dimers are required for a sexual phenotype (i.e. for a syndetic, reductional meiosis or for the successful completion of the postmeiotic mitoses). Finally, it will be assumed that some active dimers must be present if a viable pattern of apomixis is to be exhibited. How many more active than inactive dimers are required if a sexual phenotype is to be produced is a matter for further speculation. If we consider for the moment the triploid apomictic *Aaa* genotype, it is clear that this will produce dimers in the ratio of 1*AA*:4*Aa*:4*aa* if monomers associate randomly and if their rate of production is determined by the ratio of *A* to *a* genes. This gives a ratio of 5:4 of active to inactive dimers. Thus it will be assumed that this ratio must simply be exceeded in tetraploids if these are to produce sexual phenotypes. (Note that this is, at best, a 'working assumption'. The effective ratio may vary with the protein, and thus the locus, affected by the mutation and may vary if the ratio of *A*– and *a*– monomers differs from that of *A* and *a* genes.) There are three possible diploid genotypes and five tetraploid genotypes at the *A* locus. The diploid genotypes are *AA*, *Aa* and *aa*; the tetraploid genotypes are *AAAA*, *AAAa*, *AAaa*, *Aaaa* and *aaaa*. Under the assumptions being made, homozygotes for the *A* allele are obligately sexual and homozygotes for the *a* allele are inviable. Consequently, attention will be focused on the dimers produced by heterozygous genotypes.

In these, the ratio of active to inactive dimers will depend on their method of construction. Different methods of construction are described in Table 5.4. The simplest scheme is to assume, as for the triploid case just described, that the rate of production of the different monomers is determined by the ratio of *A* to *a* genes and that the monomers pair randomly. Under these conditions the heterozygous genotypes will produce dimers and phenotypes in the ratios shown in Table 5.4(a). The diploid and two of the tetraploid genotypes are sexual, but the *Aaaa* genotype is asexual.

This pattern of dimer construction is only one of several potential

Table 5.4 The phenotypes generated by diploid and tetraploid heterozygous genotypes under differing conditions of chromosome association and oligomer formation. A ratio of more than 5:4 of active to inactive dimers is required to generate a sexual phenotype (S). Ratios equal to or less than this are apomictic (A). *AA* and *Aa* dimers are active, *aa* dimers inactive. The proportions of *A* and *a* monomers produced are assumed to be equivalent to the proportion of *A* and *a* genes at the affected locus

Genotype	Dimers			Active dimers : Inactive dimers			Phenotypes
	AA	Aa	aa				
(a) *Monomers come together randomly to form dimers*							
Aa	1	2	1	3	:	1	S
AAAa	9	6	1	15	:	1	S
AAaa	4	8	4	3	:	1	S
Aaaa	1	6	9	7	:	9	A
(b) *Monomers come together from most closely related chromosome pairs*							
Aa	0	1	0	1	:	0	S
AAaa	1	0	1	1	:	1	A
(c) *Monomers come together from homologous chromosomes*							
Aa	0	1	0	1	:	0	S
AAAa (chm. pairs AA and Aa)	1	1	0	2	:	0	S
AAaa (chm. pairs AA and aa	1	0	1	1	:	1	A
or Aa and Aa)	0	1	0	1	:	0	S
Aaaa	0	1	1	1	:	1	A
(d) *Chromosome associations random, dimers form according to the association realized*							
Aa	0	1	0	1	:	0	S
AAAa	1	1	0	2	:	0	S
AAaa (association achieved AA and aa	1	0	1	1	:	1	A
or Aa and Aa)	0	1	0	1	:	0	S
Aaaa	0	1	1	1	:	1	A

ones. Different patterns emerge if the relationships between the chromosomes carrying the *A*-locus are given a determining effect. There are three such relationships: chromosomes may be sisters (derived by endomitosis or non-disjunction from chromatids of the same chromosome), homologues, or homoeologues. Consider the situation in which the monomers of a dimer are preferentially obtained from the two most closely related chromosomes rather than randomly. This could occur because highly related chromosomes are more closely associated in the nucleus than more distantly related ones and dimer construction reflects this association. This pattern of association could be exhibited during interphase, with closely related chromosomes being held closely together by the positioning of their attachment plaques, or it could reflect the repositioning and especially the pairing relationships of chromosomes during meiosis. With respect to the latter, it is known that bivalent formation in allotetraploids is frequently between homologous rather than homoeologous chromosomes and that sister, rather than homologous, chromosome pairing is common in diploid organisms that habitually undergo endomitosis before meiosis to produce autotetraploid generative cells (e.g. ferns undergoing Döpp-Manton type apomixis and parthenogenetic lizards, grasshoppers and earthworms: Walker, 1979; and references in Maynard Smith, 1978: 46). Bivalent configurations may not be relevant to apomicts undergoing completely asyndetic meioses but may be relevant to those undergoing some pairing and to apospores undergoing fully syndetic reductional meioses. The phenotypic consequences of adopting these different assumptions are as follows.

First consider a diploid *Aa* heterozygote and the *AAaa* tetraploid derived from this by endomitosis, and assume that the pairs of monomers in the dimers come from the most closely related chromosomes. These are the homologous chromosomes in the diploid, producing *A-a* associations, but the sister chromosomes in the tetraploid, producing *A-A* and *a-a* associations. Table 5.4(b) shows that the diploid produces only heterozygous active dimers and is sexual but that the tetraploid produces an equal number of active and inactive dimers. This reduction from a 1:0 ratio of active to inactive dimers in the diploid to a 1:1 ratio in the tetraploid will result in the latter being unable to synthesize sufficient active dimers, thus providing the conditions necessary for the expression of the apomictic phenotype. This is a particularly interesting result as it shows that polyploidy can induce a switch to apomixis even if it does not generate a change in the ratio of mutant to wild-type genes, which is 1:1 at both ploidy levels.

Apomicts will exhibit genomes containing sister chromosomes if they have achieved polyploidy through an endomitotic event. However, polyploidy can arise by other routes. For example, tetraploid apomicts could result from the fertilization of a haploid egg of a diploid sexual by triploid pollen of an apomict. The chromosomes of the tetraploid apomictic progeny will lack sisters but will often have homologous or homoeologous partners. Thus it is pertinent to consider the situation in which the locus carrying the apomixis gene in the tetraploid is distributed between homologous and homoeologous chromosomes. Chromosome pairing here is likely to be preferentially between homologous chromosomes. But copies of a gene could be on homologous or homoeologous chromosomes. This can be illustrated by considering genotype *AAaa*. If copies of the same gene are on homologous chromosomes then *A* will be paired with *A* and *a* with *a*. But if they are on homoeologous chromosomes *A* will be paired with *a*. This will have important consequences if the dimers reflect homologous chromosome pairing. It can be seen from Table 5.4(c) that, under this assumption, the *Aa* diploid and *AAAa* tetraploid genotypes are sexual but the *Aaaa* genotype is apomictic. The phenotype of the *AAaa* tetraploid is dependent on whether the same alleles are carried by homologues (i.e. *AA* and *aa* pairing and dimer formation) or by homoeologues (i.e. *Aa* pairing and dimer formation). If they are carried by homologues an apomictic phenotype is produced but a sexual phenotype is produced if they are carried by homoeologues.

The next scheme to be considered allows associations between sister, homologous and homoeologous chromosomes to be random but for dimers to form according to the association realized. It can be seen from Table 5.4(d) that the diploid and the *AAAa* tetraploid are sexual and that the *Aaaa* tetraploid is apomictic. However, the *AAaa* tetraploid will form associations (and thus dimers) *AA* and *aa* in some generative cells, resulting in apomictic reproduction in some ovules, but it will form associations *Aa* in other generative cells, resulting in sexual reproduction in some ovules. Consequently, this genotype will generate a facultatively apomictic phenotype.

The schemes presented above show that under all situations *Aa* diploids and *AAAa* tetraploids are sexual and *Aaaa* tetraploids are apomictic. However, the phenotypes produced by *AAaa* tetraploids are susceptible to modification. This genotype will generate an apomictic, a sexual or a facultatively apomictic phenotype depending on the pattern of association between the chromosomes carrying the locus controlling apomixis and on the way in which monomers come together to form a dimer. As stated, the biochemical basis of the scheme is highly speculative. But there is no doubt that some such scheme is required as there are data available that are difficult to explain in the absence of such a scheme. The best example of these types of data is provided by Reddy and D'Cruz (1966) and D'Cruz and Reddy (1971) who investigated the control of apomixis in *Dichanthium*.

Reddy and D'Cruz (1966) describe how they pollinated tetraploid (2*n* = 40), facultatively aposporous *D. annulatum* with pollen obtained from the sexual diploids (2*n* = 20) *D. aristatum* and *D. caricosum*. The male parents were very distinct from the female parent, differing from it in both morphological and floral characteristics and in growth form. A total of 161 mature offspring were obtained, 148 of which were maternal types produced apomictically, 11 of which were sexually produced hybrids and two of which were classed as 'atypical'. One atypical plant was produced by each of the two types of crosses. They were atypical in that they resembled the male parent, and to such an extent that Reddy and D'Cruz decided that they had been produced androgenetically. That is, these two plants contained only paternal genes. Cytological analyses confirmed this and also showed these androgens to be tetraploid ($2n = 40$). Reddy and D'Cruz investigated male meioses in the sexual species and found it to be restitutional on occasion. They surmised that the tetraploid level in the androgens could have resulted from unreduced ($2n = 20$) pollen grains through one of two routes. Both involve the two male gametes released by the pollen fusing in the pollen tube. This fused cell may then have developed into an embryo which replaced the embryo developing from the egg and utilized its endosperm, or its nucleus may have entered the egg and replaced the egg nucleus. Analysis of the cytoplasm of the androgenetic plants would be required to see which of these routes was taken, as in the former scheme the embryo will have inherited its cytoplasmic organelles from the male parent whereas in the latter these would be maternally inherited. This analysis was not performed.

The formation of androgenetic plants is of considerable interest in itself, but what makes this research particularly relevant to the present discussion is that the authors later reported that the two androgens were apomictic (D'Cruz and Reddy, 1971). D'Cruz and Reddy also produced tetraploid apospores from crossing the two sexual diploid species. The results of these crossing experiments are summarized in Table 5.5, where the crosses that resulted in the production of the two androgens are numbered 3 and 4. Interestingly, when *D. carico-sum* was used as the female parent in the cross with *D. aristatum* only diploid sexual hybrids were produced (cross 1 in Table 5.5) but in the reciprocal cross (cross 2 in Table 5.5) two tetraploid hybrids were produced, at least one of which was apomictic. (It is difficult to determine the status of the progeny produced by cross 2 and given the code BC2A in Table 5.5. This only produced a single offspring and it did so sexually. However, the possibility that it was facultatively apomictic cannot be excluded.) The tetraploid hybrids resulting from cross 2 probably resulted from endomitoses of diploid genomes that are the products of fusions between meiotically reduced (haploid) eggs and male gametes. The reasoning behind this conclusion is that meiotically unreduced gametes appear to have been produced only rarely by the parents; such gametes would therefore be most likely to fuse with the much more numerous meiotically reduced gametes that the

Table 5.5 A summary of D'Cruz and Reddy's (1971) data on the results of crossing sexual and apomictic individuals of *Dichanthium*

Parents	Code	2n
Species used to produce F1		
D. annulatum	Aa	40
D. aristatum	Bs	20
D. caricosum	Cs	20

	Code of F1		F2 plants derived		2n of sexually derived F2 plants
Crosses (f × m)			sexually	apomictically	
Sexual × Sexual					
1. Cs × Bs	CB1s	20	48	0	20
	CB2s	20	67	0	20
	CB3s	20	52	0	20
2. Bs × Cs	BC1A	40	1	3	40
	BC2?	40	1	0	50
Androgenic plants obtained during crosses					
3. Aa × Bs	BBA	40	0	21	–
4. Aa × Cs	CCA	40	0	64	–

s, sexual; A, aposporous; ?, reproductive status unclear

parents produced (to give triploid hybrids) rather than with other unre-duced gametes (to give tetraploids).

Thus the differences in the phenotypes of the androgens and their male parents and of the reciprocal crosses 1 and 2 provide separate lines of evidence that a simple doubling of a genome can result in a switch from sexual to apomictic reproduction. Superficially, this appears to provide evidence that, in the absence of 'apomixis genes', apomixis can be induced by polyploidization. However, D'Cruz and Reddy (1971) make the crucial observation that the diploid sexual individuals of *D. caricosum* and *D. aristatum* are probably polyhaploid derivatives of facultatively apomictic tetraploids. Such polyhaploids could have resulted from the occasional parthenogenetic development of meiotically reduced (diploid) eggs. The importance of this obser-vation is that it demonstrates that the sexual diploids could easily have been carriers of apomixis genes, but in ratios with wild-type genes that were not conducive to their phenotypic expression. Under the terms of the model described in the previous section they would exhibit genotype *Aa* (genotype 2, Table 5.1). This diploid genotype will be given irrespective of whether the putative tetraploid ancestor exhibited genotype *AAaa* or *Aaaa*, as each is capable of generating *Aa* diploid gametes. Thus the androgens would exhibit genotype *AAaa*. So too would most of the tetraploids resulting from crossing the two sexual diploids, as this cross will result in the viable genotypes *AA* and *Aa* and the inviable genotype *aa* in a 1:2:1 ratio. A duplication of the viable genotypes will produce twice as many *AAaa* tetraploids as *AAAA* tetraploids.

D'Cruz and Reddy's results confirm the need, described above, to consider the possibility that the actual number of wild-type and mutant genes as well as their ratio may help to determine whether a sexual or an apomictic phenotype is produced, as the ratio of genes in the diploid sexual species and hybrids must have been the same as that in their tetraploid apomictic derivatives. As these tetraploids resulted from endomitosis they would have had genomes comprising pairs of sister chromosomes. Under the assumptions made in the determination of the results depicted in Table 5.4(b) (that sister chromosomes preferentially associate and that monomers come together to form dimers according to the pattern of chromosome associations) genotype *AAaa*, but not genotype *Aa*, produces an apomictic phenotype. D'Cruz and Reddy's (1971) results (summarized in Table 5.5) are consistent with this, and consequently with the model for a single locus control of apomixis developed earlier in this chapter. The authors, however, offer a different explanation, suggesting that their data indicate that there is a delicate balance between apomixis and sexuality and that the genes controlling these forms of reproduction are not allelic. However, they do not elaborate on this and it is not clear how they arrive at this conclusion.

The accumulation of surplus apomixis genes and its effects on the control of apomixis

Although there have been a number of investigations into the control of apomixis few have resulted in the publication of methods or data that are in a form that can be subjected to reanalysis. The data set is often too small to enable statistical analyses to be performed, or the methods and materials used in the experiments are inadequately explained or described. For example, Harlan *et al.* (1964) examined the variability of F2 families produced from hybrids formed from crosses between sexual and aposporous plants of *Dichanthium* and of the closely related genus *Bothriochloa*. They used these measures of variability, rather than direct examination of the ovules of the hybrid F1 generation, to determine the reproductive behaviour of the hybrid F1. Families that were most variable were deemed to have been produced sexually whereas those that were less variable were deemed to have been produced aposporously. However, the authors do not describe the nature or variability of the characters used in this determination. Nor do they indicate whether the conditions under which sexual reproduction will have occurred favoured selfing or outcrossing. As *Dichanthium* apospores are facultatively so, both sexual and asexual families exposed to the same pollen source could exhibit the same range of segregants for a character, as apospores as well as sexual individuals will give rise to progeny by sexual reproduction. The variance of a character among the progeny of a plant will not depend simply on the range of the character but will be affected both by the ratio of aposporously to sexually produced progeny and by the ratio of self-fertilized to outcrossed progeny among the latter. The lack of information provided by the authors on these aspects of their investigation means that their data cannot be practicably reanalysed.

Fortunately, some published data are suitable for analysis. Two sets, relating to *Taraxacum* and *Dichanthium*, have already been reanalysed and have been found to be in agreement with the assumptions of the single locus model proposed for the control of apomixis. There are two other suitable data sets available. One is that published on *Cenchrus ciliaris* (*Pennisetum ciliare*) by Taliaferro and Bashaw (1966). The other is provided by Hanna *et al.* (1973) for *Panicum maximum*. However, neither of these sets, at first glance, meets the requirements of the model. Nevertheless, it will be demonstrated in this and the following subsection that they can be encompassed by the model if one or two reasonable assumptions are made.

The data provided by the investigations (Hanna *et al.*, 1973) into the control of apomixis in *Panicum maximum* will be reanalysed first. But before this exercise is initiated I wish to discuss a problem that will arise on occasion and that concerns the interpretation of data. This is a necessary prelude to the analysis of the *Panicum* data.

Within the constraints of a single locus control for apomixis in preadapted taxa, the ratio of sexual and apomictic progeny expected from a cross can be easily calculated. This has been done for the tetraploid case using a number of different assumptions. These ratios are given in Table 5.4. However, although the model presented argues for a single locus control, it explicitly states that the locus involved in any particular apomictic lineage will be only one of many that could be involved. This is a complication that should not be ignored as it provides the means by which a lineage could easily accumulate apomixis genes at more than one locus, even though it only needs to acquire them at a single locus in order to generate an apomictic phenotype. Such multiple accumulations could be achieved through mutation, through sexual reproduction between facultatively apomictic lineages that have acquired apomixis genes at different loci, or through sexual reproduction between facultatively and obligately apomictic lineages (with the obligately apomictic lineage acting as the male parent) which differ in the identity of the apomixis-conferring locus. With respect to sex-mediated gene exchange between apomictic lineages, two such sympatric and compatible lineages would, within a few generations, generate new apomictic lineages containing both sets of apomixis genes. An example of this is given for the tetraploid case in Fig. 5.1, and the full range of genotypes and phenotypes that could be achieved after several generations is given in Table 5.6 for crosses between two tetraploid facultatively apomictic lineages exhibiting apomictic genotypes at different loci. In both cases, a combination at a locus of two wild-type and two apomixis genes is considered to be sufficient to generate an apomictic phenotype. The point I wish to make here is that it may be possible to demonstrate that a lineage carries apomixis genes at more than one locus, but this does not provide conclusive evidence that the evolution of apomixis in the lineage required the accumulation of mutations at more than one locus. This issue has been raised previously by Stebbins (1950) who pointed out that crosses between apomictic lineages (using a facultative apomict as the maternal parent) that express different combinations of apomixis genes often result in a preponderance of sexual progeny. This result is to be expected if apomixis is expressed through

Lineages:		1		2

Genotype at the A and B loci : AAaa BBBB   AAAA BBbb

Gametes: 1AABB : 2AaBB : 1aaBB   1AABB : 2AABb : 1AAbb

Matrix of progeny genotypes

	1AABB	2AaBB	1aaBB
1AABB	1AAAA BBBB (S)	2AAAa BBBB (S)	1AAaa BBBB (A)
2AABb	2AAAA BBBb (S)	4 AAAa BBBb (S)	2AAaa BBBb (A)
1AAbb	1AAAA BBbb (A)	2AAAa BBbb (A)	1AAaa BBbb (A*)

Figure 5.1 Progeny genotypes and phenotypes resulting from a cross between two facultatively apomictic lineages. Lineage 1 is apomictic because of the balance of apomixis and wild-type genes at the A-locus. It is homozygous for wild-type B-genes at the B-locus. Lineage 2 is apomictic because of the balance of apomixis and wild-type genes at the B-locus. It is homozygous for wild-type genes at the A-locus. A 9:7 ratio of sexual (S) to apomictic (A) progeny is given. Both the A and the B loci in progeny type (A*) each carry combinations of wild-type and apomixis genes that would generate an apomictic phenotype. Three of the remaining six apomictic types carry a suitable combination at the A-locus but not at the B-locus, and three carry a suitable combination at the B-locus but not at the A-locus.

Table 5.6 The phenotypes produced by genotypes containing apomixis genes at none, either or both of two loci. The wild-type genes (coding for sexuality) are *A* and *B*. The apomixis genes are *a* and *b*. A homozygote for an apomixis gene is inviable. Otherwise an apomictic phenotype results if either locus carries two or more apomixis genes

Genotype	Phenotype
AAAA BBBB	sexual
AAAA BBBb	sexual
AAAA BBbb	apomictic
AAAA Bbbb	apomictic
AAAA bbbb	inviable
AAAa BBBB	sexual
AAAa BBBb	sexual
AAAa BBbb	apomictic
AAAa Bbbb	apomictic
AAAa bbbb	inviable
AAaa BBBB	apomictic
AAaa BBBb	apomictic
AAaa BBbb	apomictic
AAaa Bbbb	apomictic
AAaa bbbb	inviable
Aaaa BBBB	apomictic
Aaaa BBBb	apomictic
Aaaa BBbb	apomictic
Aaaa Bbbb	apomictic
Aaaa bbbb	inviable
aaaa BBBB	inviable
aaaa BBBb	inviable
aaaa BBbb	inviable
aaaa Bbbb	inviable
aaaa bbbb	inviable

a dosage effect of apomixis genes at a single locus. This can be seen in Fig. 5.1, where the majority of progeny produced by the cross depicted are phenotypically sexual.

The investigations by Hanna *et al.* (1973) into the control of apomixis in aposporous *Panicum maximum* generated data that can be interpreted as demonstrating the existence of surplus genes for apomixis. Apomixis in this species is usually obligate although some clones appear to produce sexual individuals on occasion (Warmke, 1954; Bogdan, 1963). The authors utilized 158 different accessions of *Panicum maximum* obtained from an agricultural research station. They identified 295 clones from these, two of which were found to be obligately sexual. In all, 18 plants were identified as being sexual. Ten of these were exposed to pollen from apomictic clones and to self-pollen, resulting in the production of an F1 comprising both hybrid progeny and progeny produced by selfing. Of the 49 hybrids that were examined cytologically 21 were determined as sexual and 28 as apomictic. However, the segregation ratios of individual crosses were not reported resulting in this part of their investigation being unsuitable for reanalysis. Chromosome counts of some of the sexual plants indicated that they were all tetraploid ($2n = 32$). The ploidy levels of the apomicts are not reported but Savidan (1980) states that they are typically tetraploid in this genus.

Four of the sexual F1 hybrids were allowed to self-fertilize. These produced a mixture of sexual and apomictic progeny, with the numbers produced being provided for each parent. The following discussion will concentrate on these data (summarized in Table 5.7(a)).

Table 5.7 (a) Summary of the data obtained by Hanna *et al.* (1973) by selfing four F1 sexual plants (a-d) obtained from a population containing aposporous and sexual individuals of *Panicum maximum*. The 11:5 segregation ratio is that predicted by Hanna *et al.*'s model. The 9:7 and 3:1 ratios are those predicted in the text. The derivation of these latter ratios is shown in (b); the genotypes are those expected if it is assumed that apospory is controlled at either of two loci

(a)

	Observed number of progeny		Observed ratio	X^2 on 11:5 ratio	X^2 on 9:7 ratio	X^2 on 3:1 ratio
Plant	Sexual	Apomictic				
a	12	5	2.4:1	0.03	1.42	0.18
b	77	33	2.3:1	0.08	8.44	1.47
c	10	9	1.1.1	2.29	0.12	5.07
d	17	7	2.4:1	0.05	2.11	0.22

(b)
Selfing the sexual phenotypes derived in the text gives:
AAAA BBBB × AAAA BBBB = 1 sexual:0 aposporous
AAAA BBBb × AAAA BBBb = 3 sexual:1 aposporous
AAAa BBBB × AAAa BBBB = 3 sexual:1 aposporous
AAAa BBBb × AAAa BBBb = 9 sexual:7 aposporous

Hanna *et al.* (1973) noted that the segregation ratios obtained from selfing the four sexual hybrids showed a good fit to an 11:5 ratio. They concluded from this that sexuality is dominant to apospory and that it is controlled by at least two loci and conditioned by a dosage effect of two or more dominant alleles. They named these two loci (which must segregate independently to generate the predicted ratio) as A and B and listed the range of combined genotypes at these as comprising

AABB, AABb, AAbb, AaBB, AaBb, aaBB, Aabb, aaBb and *aabb*. They argued that a sexual phenotype results if the genotype contains two or more dominant (i.e. *A* or *B*) alleles. Whether the two dominant alleles are at the same locus or at different loci does not matter. Thus the first six genotypes listed above generate sexual phenotypes, the remainder apomictic phenotypes. They concluded that the segregation ratio observed among selfed progeny will result from the selfing of individuals with the sexual *AaBb* genome.

There are several problems with this analysis. One is that it is unreasonable to assume that the sexual parents should all be *AaBb*. Under the genetic scheme proposed by the authors there are six sexual tetraploid genotypes. As the sexual F1 hybrids resulted from open pollination it is reasonable to assume that five of the six sexual genotypes should be represented (genotype *AABB* will be absent as *AB* pollen cannot be produced by apomicts). If all are at equal frequency the probability of randomly selecting four that are each *AaBb* is less than 0.2%. It is more likely that the sexual plants exhibited more than one genotype. Thus a constant ratio of sexual to apomictic offspring should not be expected on selfing these plants.

A further problem is that the model is described in terms of a diploid genome even though the plants were tetraploid. Presumably, the authors assume disomic rather than tetrasomic inheritance but this is not made clear. If so, this is an added complication and it is worth considering whether there is a simpler interpretation of the data. The requirement of Hanna *et al.* (1973) that a certain number of the alleles present at the two affected loci must be dominant (i.e. *A* or *B* alleles), but that it does not matter whether these reside at one locus or are distributed between both is unusual. Nevertheless, the segregation ratios do seem to require some interaction between loci as none of the various forms of the one locus model developed from the *Dichanthium* data generate sexual and apomictic progeny in a ratio that fits the *Panicum* data. However, as argued above, a requirement to consider two loci does not offer conclusive evidence that apomixis requires this level of control.

An important point to take into consideration is that the original 158 accessions used by the authors were obtained from an agricultural research station and not from a natural population. It is probable that these accessions had been collected over a wide area and that they represented several apomictic lineages. It is reasonable to assume that there may have been variation among these in the identity of the locus or loci controlling apospory. The two sexual lines obtained from this artificial population may well have resulted from crosses between apomictic lines, some of which, as indicated earlier, may retain some capacity for sexual reproduction. Alternatively, they may have resulted from crosses between sexual lineages and apomictic lineages (acting as male parents). Consequently, the sexual accessions used in this study may each have acquired apomixis genes at more than a single locus. Thus, even if apomixis in this species need involve only a single locus control, evidence for this will be compromised by the presence of these surplus genes.

This being the case, it is useful to consider whether the four gene combinations listed in Table 5.6 (and derived in Fig. 5.1) that generate sexual phenotypes will, when selfed, generate sexual and aposporous progeny in ratios that are not significantly different from those observed by Hanna *et al.* If this can be demonstrated then it cannot be ruled out that a single locus control for apomixis is a possibility for *Panicum maximum*. The expected ratios are shown in Table 5.7(b). They are (sexual:aposporous) 1:0, 3:1 and 9:7. The data of Hanna *et al.* are summarized in Table 5.7(a). It can be seen from this that two of the expected ratios (3:1 and 9:7) are not significantly different from the observed ratios for particular plants. Plant b deviates significantly from a 9:7 ratio but not from a 3:1 ratio, and plant c deviates significantly from a 3:1 ratio but not from a 9:7 ratio. Plants a and d do not differ significantly from either ratio. The ratios from all four plants agree with the 11:5 ratio of Hanna *et al.* (1973). Thus it is worth re-emphasizing that the 11:5 ratio is that predicted if two assumptions are made that I believe are rather implausible – that the four selfed hybrids share the same genotype at the loci controlling apomixis, and that apomixis is disomically rather than tetrasomically inherited.

This reanalysis does not offer proof that apomixis in *Panicum maximum* requires only a single locus control. However, it does show that the data provided by Hanna *et al.* (1973) are compatible with this type of control if the presence of surplus genes for apomixis at other loci are acknowledged. The model proposed in this chapter allows for the easy accumulation of surplus genes. Thus the *Panicum* data fit comfortably within the model.

Apomixis and the re-emergence of sexuality by chromosome mutation in tetraploids

The previous discussion demonstrates one potential source of confusion which should be considered when data are being analysed – that of the accumulation of surplus genes for apomixis by a lineage. There is another. This concerns the effect of the loss or gain through non-disjunction or mutation of copies of the wild-type or mutant genes at the locus controlling apomixis. This problem has already been discussed for the triploid case, where it was clear that switches between sexual and apomictic phenotypes accompanied such changes in *Taraxacum*. The purpose here is to extend the discussion of this phenomenon to the tetraploid case. Data relevant to this discussion are provided by Taliaferro and Bashaw (1966) for *Cenchrus ciliaris* (*Pennisetum ciliare*).

Apomicts of this taxon are obligately aposporous. Taliaferro and Bashaw obtained a single sexual individual which as shown in earlier work (Bashaw, 1962, reported in Taliaferro and Bashaw, 1966) produced a mixture of obligately sexual and obligately apomictic progeny. Clones were obtained from this sexual individual by vegetative multiplication and were typified SB. The SB type was selfed and was also used as a female in crosses with two apomictic clones which were typified BB and CB. Each of the three types was tetraploid ($2n=36$). When used as a female in crosses with the apomicts the SB type was emasculated and hand pollinated. However, emasculation does not appear to have completely prevented the SB type from self-fertilizing, as the authors report that the mean percentage seed set for 25 inflorescences that were emasculated but not hand pollinated was 2.6%, against a mean of 27% for several inflorescences that were both emasculated and hand pollinated. The method of reproduction of the progeny of these crosses was determined by observation of mature embryo sacs and/or by an examination of progeny variability. Briefly, the first selfed (SB × SB) generation (S1) had its mode of reproduction determined by embryo sac analysis. Seeds obtained from this generation were grown to produce an S2 generation and it was noted that the S2 from sexual S1 plants were much more variable morphologically than those from apomictic S1 plants. Consequently, the majority of the F1 plants obtained from crossing SB with BB and CB were classified as sexual or apomictic by observation of 10- to 20-plant F2 progeny rows. The results of these experiments are summarized in Table 5.8.

The authors argue that the segregation ratios (sexual:apomictic) obtained (Table 5.8) indicate that two independently assorting loci control the expression of the aposporous phenotype. They give the genotype at these loci in the SB sexual as *AaBb*, and argue that the *B* gene conditions sexuality and is epistatic to the *A* gene which conditions apomixis. Any genotype that contains at least one *B* allele will generate a sexual phenotype, whereas any that lacks this allele but which contains at least one *A* allele will generate an apomictic phenotype. Surprisingly, the authors assume that the double recessive homozygote (*aabb*) will generate a sexual phenotype even though it lacks the gene that conditions sexuality. They argue that a sexual phenotype will result from this genotype because it lacks the dominant gene conditioning asexuality. Under the authors' model, selfing of SB will generate progeny in the ratio of 13 sexual to 3 apomictic. They further argue that the approximately 5:3 (sexual:apomictic) ratio obtained by

Table 5.8 The observed number of sexual and apomictic plants produced by Taliaferro and Bashaw (1966) from crosses involving a sexual (SB) and two aposporous (BB and CB) clones of *Cenchrus ciliaris*, and the expected number assuming a 13:3 or a 23:4 segregation (sexual:aposporous) in the SB × SB cross and a 5:3 or a 2:1 segregation in crosses between SB and the apospores

Year	Observed number of sexual and apomictic plants		Expected number on a 13:3 hypothesis		χ^2	Expected number on a 23:4 hypothesis		χ^2
	Sexual	Apomictic	Sexual	Apomictic		Sexual	Apomictic	
Cross = SB × SB								
1962	315	68	311.2[a]	71.8[a]	0.25	326.3	56.7	2.64
1963	130	15	117.8	27.2	6.72**	123.5	21.5	2.31
1965	72	14	69.9	16.1	0.34	73.3	12.7	0.16
Pooled	517	97	498.9	115.1	3.50	523.0	91.0	0.47

Cross	Observed number of sexual and apomictic plants		Expected number on a 5:3 hypothesis		χ^2	Expected number on a 2:1 hypothesis		χ^2
	Sexual	Apomictic	Sexual	Apomictic		Sexual	Apomictic	
SB×CB	373	202	359.4	215.6	1.39	383.3	191.7	0.83
SB×BB	108	83	119.4	71.6	2.89	127.3	63.7	8.77**

** Significant χ^2 (*P* > 0.05)
[a] Corrections of a mistake in Table 1 of Taliaferro and Bashaw (1966)

crossing SB to the apomicts (Table 5.8) indicates that the genotypes of the apomicts must be *Aabb*. Finally, they argue that the 5:3 ratio indicates that the F1's are probably back-crosses and they suggest that sexuality in the SB clone may have been achieved through mutation at the *B*-locus.

This model, like that of Hanna *et al.* (1973) just described for *Panicum maximum*, is problematic in that it offers only two alleles per locus for a tetraploid genome and is thus appearing to assume diploidization. In fact, the data can be interpreted as indicating a single locus control of apospory. However, in order to do so it is necessary to assume that the sexual clone arose from an apomict following a deletion mutation that caused the loss of one copy of the apomict's apomixis gene. Using the notation developed in the model (*A* = wild type gene; *a* = apomixis gene), consider the situation in which the *AAaa* genotype generates a sexual phenotype and the *Aaaa* genotype generates an aposporous phenotype (as would occur under several of the schemes described in Table 5.4). If the apospore lost one of its *a* alleles it would exhibit genotype *Aaa-*. The change in the ratio of *A* to *a* genes will be sufficient to generate a sexual phenotype. (For example, with the random pairing of monomers into dimers, the ratio of active to inactive dimers would change from a 7:9 excess of inactive dimers to a 5:4 excess of active dimers, which is at the boundary given earlier at which it was proposed that the switch from apomictic to sexual reproduction occurs.)

If it is assumed that the sexual SB clone has the genotype Aaa-, selfing will generate progeny in the ratio of 23 sexual:4 apomictic:9 inviable (sexual = 4AAaa, 4AAa-, 10Aaa-, 1AA--, 4Aa--; apomictic = 4Aaaa; inviable = 1aaaa, 4aaa-, 4aa--). This 23:4 ratio among viable progeny fits the data better than the 13:3 ratio predicted by the authors' two-locus model (Table 5.8). Crossing genotype Aaa- to apomictic genotype $Aaaa$ generates viable progeny in the ratio of 2 sexual:1 apomict. This ratio fits the data for the SB × CB cross better than the 5:3 ratio predicted by the authors' model although it does not agree with the results of the smaller sample from the SB × BB cross (Table 5.8). All in all, the simpler one-locus model proposed is at least as convincing as the authors' two-locus model.

The failure of the data from the SB × BB cross to fit the one-locus model gives no great cause for concern. As described earlier, the progeny of the crosses between the sexual clone and the apomicts were classified as being sexual or apomictic by examining the patterns of segregation among a small number (10 – 20) of F2 progeny. This method of determining reproductive behaviour is the same as that used by Harlan *et al.* (1964) for *Dichanthium* and criticized earlier. However, it probably generates less error when applied to *Cenchrus* than when applied to *Dichanthium* as the former is obligately apomictic whereas the latter is facultatively so. The progeny of an obligate apomict will be much less variable than those of a facultative one (as the progeny of the latter will include a mixture of sexually and aposporously derived individuals) and very much less variable than that of a sexual plant. However, an error rate of only 3.7% is all that is required to turn the SB × BB cross into a significant result (i.e. had the observed results been 115 sexuals and 76 apomicts instead of 108 sexuals and 83 apomicts they would have fitted a 2:1 ratio). The fact that emasculation of the SB type did not prevent some selfing leaves open the possibility that some progeny believed to have resulted from crossing the SB mother to apomictic males resulted instead from selfing of SB. Such progeny would be maternal in appearance and, because of this, could be misclassified as having been aposporously derived. Because of this, and because of the small numbers of F2 progeny involved in the determination of reproductive behaviour, rows with an excess of selfed progeny may have been mistakenly classified as being derived from an apomictic rather than a sexual F1 mother. This could have resulted in the significant departure from expected in the ratio of sexual to apomictic progeny in the SB × BB cross. It is interesting to note that although the ratio in the SB × CB cross was not significantly different from 2:1 it was nevertheless lower, being

1.85:1. A deviation in this direction is to be expected if there is a possibility of selfed progeny being misclassified, because of their maternal appearance, as having been derived apomictically.

5.6 DISCUSSION

The model proposed in section 5.4 and developed in section 5.5 for the control of generative and aposporous apomixis describes a single locus control of these processes. In section 5.5, it was shown that data from four apomictic taxa can be reinterpreted in terms of the model. Some of these data do not fit easily within its confines and several assumptions have had to be made to accommodate them. These difficulties may be viewed as providing evidence that the model lacks robustness. However, its form is such that difficulties like these are to be expected. Testing data against a one locus model is straightforward if the identity of the determining locus is fixed. But the identity of the locus in the model is fluid, differing between lineages. This is simply because it is valid to argue that mutations at any one of a number of loci involved in the control of meiosis may generate apomictic phenotypes. Because of this, different apomictic lineages can be expected to vary in the location of the locus responsible for their reproductive mode. This in itself will not make data analysis difficult. But difficulties of interpretation will arise whenever lineages acquire surplus apomixis genes at other loci; this may arise due to mutation or, when apomixis is facultative, to gene flow between lineages. The fact that the proposed control is simple adds further to the likelihood that difficulties of interpretation will arise, because it will be easily upset by subsequent mutation. Thus, it has been shown how either the acquisition of only a single additional copy of the wild-type gene or the loss of only a single copy of a mutant gene can result in the re-emergence of a sexual phenotype in an otherwise apomictic lineage. It is worth bearing this point in mind whenever a data set obtained from experiments involving rare sexual forms is subjected to analysis. If the origin of these forms is unknown, the possibility should be considered that they have arisen by mutation from apomictic forms and that they may consequently carry genes for apomixis that are not, however, phenotypically expressed.

The advantages of this model over the others that have been proposed in the literature are its simplicity, its ability to explain facultative as well as obligate apomixis, its ability to explain both aposporous and generative apomixis, its ability to explain the variation observed within each of these forms, and its ability to interpret both the association between apomixis and polyploidy, and the segregation patterns observed in crosses involving apomicts, in terms of gene dosage effects.

The advantages of simplicity have been stressed repeatedly but it is nevertheless worth stressing them once more. An evolutionary change that would prove advantageous and that can be achieved simply by one group of taxa but with some difficulty by another is likely to be achieved more frequently in the former than in the latter. One implication of this is that the demonstration that apomicts have evolved apomixis very easily, through a single mutational step, cannot be used as evidence that it is always an easy state to achieve. To do so is as erroneous as using the fact, say, that many novels have been published to conclude that novels are easy to get published. In both cases a value judgement is being made on the basis only of the number of successes. A balanced judgement requires additional knowledge – that of the number of failures. Thus it is worth stating again that although the models described in this chapter depict the evolution of apomixis as a rather simple affair, I am not claiming here that apomixis is always easy to evolve. What I am claiming is that taxa which have given rise to apomictic lineages have done so easily. Its evolution in other taxa will be much more difficult, indeed so much more so that most, if not all, have been unable to evolve it and consequently must continue to reproduce sexually, or not at all.

This chapter has dealt with a number of complex issues and these are briefly summarized before moving on to the next chapter. The main points are:

1. Views on the control of parthenogenetic apomixis have changed over the years. Initially, it was considered likely that polyploidization and/or hybridization were sufficient to provide the conditions necessary for the expression of the apomictic phenotype. This opinion has not yet fully succumbed but it is now usually accepted that apomixis requires the accumulation of apomixis genes.

2. Views differ regarding the number and primary purpose of these genes and their relationship to wild-type genes coding for sexual reproduction. Most models incorporate two or more apomixis genes at two or more loci. Some models do not specify the action of these genes, but others consider that some are responsible for the meiotic irregularities associated with apomixis, some with the development of aposporous embryo sacs, some with the avoidance of fertilization and some with parthenogenesis. The models differ in their opinions about whether the apomixis genes are dominant or recessive to the wild-type genes.

3. The model proposed here is based on observations of ontogenetic, developmental and mutational events in sexual organisms – primar-

ily sexual insects and flowering plants. These show that parthenogenesis is often an innate capacity and they indicate that the meiotic irregularities associated with parthenogenetic apomixis may
be associated with a family of meiotic mutations. If so, it is likely
that only a single meiotic mutation, at any one of a number of loci,
is sufficient to generate the apomictic-like phenotype. A likely
consequence of these mutations is a reduction in the time taken
to produce a mature egg. This reduction may allow eggs to pass
the stage at which they can be fertilized before male gametes are
able to gain access to them, leading to the phenomena of precocious
oogenesis and precocious embryony described in Chapter 3. In
sexual organisms, some meiotic mutations lead to the types of
restitutional or mitotized meioses typical of generative apomicts,
but some cause the degeneration of the generative cell or of its
mitotic products – a phenomenon that is characteristic of aposporous apomicts. With respect to aposporous apomicts, it is argued
that factors escaping from the degenerating cells will stimulate
neighbouring somatic cells to differentiate as embryo sacs. Thus it
can be argued that a mutation at a single locus involved in the
control of meiosis is sufficient to generate an apomictic phenotype
in some taxa.

4. Other taxa may not be pre-adapted for parthenogenetic apomixis
in this way. In these, more than a single mutation will be necessary
for the evolution of apomixis. These genes will need to be acquired
and phenotypically expressed simultaneously, or almost so, as they
are advantageous under these circumstances but deleterious otherwise. The probability of achieving this pattern of acquisition and
expression is remote. This may account for some of the variation
in the taxonomic distribution of apomixis, although some will be
due to other factors, including ecological factors.

5. In the same way that taxa may differ in the ease with which they
could acquire apomixis, taxa which are able to acquire it easily
may differ in the ease with which they can acquire particular forms.
A taxon which is pre-adapted to acquire generative apomixis is not
necessarily pre-adapted to acquire aposporous apomixis (and *vice-
versa*). Thus variation in the distribution of generative and aposporous apomixis is to be expected in taxa in which apomixis is
common.

6. In sexual organisms, the meiotic mutations that have been proposed here as being responsible for the evolution of apomixis in
pre-adapted taxa are recessive to wild-type genes coding for sexual
reproduction. Many also reside at loci which play a part in the
control of mitosis, and their presence can disrupt this process leading to inviable patterns of growth and development. It is argued that

apomicts have circumvented this problem by achieving phenotypic dominance of the mutant gene in generative cells but phenotypic dominance of the wild type gene in vegetative cells. It is proposed that the dominance of the mutation in generative cells is achieved through a dosage effect, with excess copies of the mutant gene dominating fewer copies of the wild-type gene. The easiest way to achieve a ratio of wild-type to apomixis genes conducive to the generation of an apomictic phenotype is through polyploidy, thus explaining why most parthenogenetic apomicts are polyploid. It is proposed that the dominance of the wild-type gene in vegetative cells is achieved because the dominance relationship between the two types of gene is affected by the environment as well as by the number of copies of each type, with the environment of the vegetative cells favouring the phenotypic dominance of the wild-type gene, and the very different environment of the generative cell favouring the phenotypic dominance of the mutant gene. However, different generative cells of an individual can experience different environments. In some cases these differences may be so great that the dominance relationship between the two types of gene will vary between ovules, resulting in facultative apomixis.

7. Although apomixis may result from the acquisition of a mutation at a single locus in pre-adapted taxa, many of these may exhibit apomixis genes at more than one locus. This can happen in three ways: first, by mutation; second, by gene exchange between apomictic lineages that have acquired different apomixis genes at different loci; this may be common in populations of compatible facultative apomicts; third, by the accumulation of modifier genes after the evolution of apomixis. The latter will often be required because the initial apomictic phenotype may be suboptimal. This may be due, for example, to egg maturation in flowering plants being too precocious or not sufficiently precocious with respect to endosperm development. Such accumulations will complicate genetic analysis of the control of apomixis, as a distinction must be made between genes that are necessary for the expression (evolution) of an apomictic phenotype, and those that are either surplus to requirements or are concerned with improving an already viable apomictic phenotype.

Chapter 6

At the Court of the Red Queen

6.1 INTRODUCTION

Explanations for the maintenance of sexual reproduction have consistently been sought (e.g. Darwin, 1862; Fisher, 1930; Muller, 1932; Gustaffson, 1946–47), and their acquisition has been a major goal of evolutionary biology for the past two decades (e.g. Ghiselin, 1974; Williams, 1975; Maynard Smith, 1978; Bell, 1982; Stearns, 1987a; Michod and Levin, 1988). And yet, as Stearns (1987b:26) succinctly puts it: 'No one has yet given a convincing, single-generation, micro-evolutionary and experimental demonstration of the advantages of sex, which must nevertheless exist.' This statement can be modified to take into account the existence of a single-generation, microevolutionary advantage of sex in homosporous plants (Mogie, 1990; and Chapter 3), although it still holds true for animals and heterosporous plants.

This chapter, will look at two aspects of this problem. The first concerns the role of genetic recombination through crossing-over (i.e. chiasma formation) in sexually reproducing organisms. This is the subject matter of Sections 6.2 and 6.3. There is a considerable literature on this, a comprehensive review of which can be found for example in the stimulating collection of essays in Michod and Levin (1988). Here I will concentrate on issues that have been generally overlooked, but which I feel have an important contribution to make to the debate. One example is the maintenance of crossing-over during sexual reproduction, but its absence during asexual reproduction, in facultative apomicts. Both the sexually and asexually produced progeny of a facultative apomict are dispersed into the same environment at the same time to face the same range of adversaries and adversities, and yet there is no firm evidence that the sexually produced component has a generation-by-generation ecological edge over the asexual one. This must cast a shadow over hypotheses that propose that the primary role of crossing-over in sexual reproduction is to provide at least some offspring with the genetic wherewithal to overcome the ecological pitfalls that mark the route to reproductive success. Against the background of these problems I will argue that the primary role of crossing-over in many taxa may be to generate a genetic environment that is suitable for the smooth running and completion of a reductional meiosis. I do not have absolute confidence in this argument, as it is based on assumptions that are difficult to verify, but I offer it for, although it lacks a strong base, other aspects of its structure are more robust.

The second issue to be considered is the evolutionary potential of asexual organisms. There are reasons to believe that the capacity for genetic change may be affected by polyploidy, by hybridization and, in apomicts, by the acquisition of the meiotic mutations that are adopted in Chapter 5 as apomixis genes. These possibilities will be investigated.

6.2   GENETIC RECOMBINATION AND OFFSPRING FITNESS

There are several potential costs of sex. Lewis (1983, 1987) lists three general ones and two that he argues are specifically related to sex in anisogamous organisms. The general costs result from recombination, from cellular-mechanical events associated with sexual reproduction, and from fertilization. The additional costs associated with anisogamy result from genome dilution and from sexual selection. The cost of recombination is that of a reduction in fitness due to the disruption of favourable gene combinations. The cellular-mechanical costs encompass any reduction in fitness due to the loss in cells of synthetic potential resulting from delays due to the time taken for a cell to complete meiosis (which is much longer than the time taken to complete two mitoses), or from delays in achieving syngamy or karyogamy. The costs of fertilization encompass any reductions in fitness due to the difficulty of bringing two gametes together from different organisms. They include reductions in fitness due to any delay or prevention of reproduction that may occur because of the unavailability of mates, or because of any increased exposure to predators, parasites or diseases that mating behaviour may generate. The cost of genome dilution is Lewis' (1987) preferred way of describing the cost of sex that results from the pattern of egg or meiospore production. However, contrary to Lewis' argument, it is not a cost associated with anisogamy *per se* but with outcrossing in a subset (albeit a very large and important one) of anisogamous organisms, namely, animals and heterosporous plants. It is not a cost associated with self-fertilization in this subset or with sexual reproduction in homosporous plants, as explained in Chapter 3. Finally, the costs of sexual selection include any penalties which the environment extracts 'as a result of phenotypic manifestations of sex that are established and maintained by sexual selection' (Lewis, 1987). They include costs of increased vulnerability and loss of time associated with sexual competition, and costs associated with the phenotypic specializations the males and females of a species may exhibit as a result of sexual selection.

The challenge that has exercised the minds of evolutionary biologists is to explain why the majority of animals and heterosporous plants reproduce sexually despite the costs of sex. However, it is becoming clear that sexual reproduction may be widespread simply because many taxa have not been able to evolve asexual reproduction. Sexual reproduction exists in these because of an absence of a challenge from asexual reproduction rather than despite such a challenge. This seems to be the case in bryophytes, gymnosperms and mammals and, as argued in Chapter 5, may typically be the case in any taxon in which the evolution of asexual reproduction will require the simultaneous acquisition and phenotypic expression of several mutations.

Nevertheless, it is probable that sexual reproduction is maintained in some animals and heterosporous plants despite real challenges by asexual reproduction, as it is clear (Chapter 5) that asexual reproduction may be able to emerge easily in some taxa, by the acquisition of only a single mutation. Thus, it is reasonable to assume that asexual mutants have arisen in some extant sexual taxa but have failed to become established. This can only occur if the potential costs of sex outlined above can be overcome.

Some of the potential costs of sex may not apply to some organisms, or may be much less severe for some organisms than for others. For example, organisms in which the time required for prereproductive development is greater than the time required for meiosis need not experience a delay in reproduction due to meiosis as long as these divisions are initiated early enough in development (Lewis, 1987). This will be the case in most multicellular organisms. Similarly, the costs of sexual selection may be experienced primarily by animals, as sexual selection may not operate to any great extent in plants.

If costs associated with sexual reproduction cannot be avoided then they must be overcome if sex is to be maintained. Costs associated with outcrossing will be reduced if, for example, paternal investment in the young enables sexually produced offspring to survive better than asexually produced offspring. They will also be reduced if sexual females are more fecund than related asexual females. This appears to be the case in some insect taxa, where asexual females have, on average, only 67% of the fecundity of sexual females (Lamb and

Willey, 1979). But it does not appear to be generally the case in flowering plants, where sexual forms can be less fecund than asexual forms because of pollen limitation (Bierzychudek, 1981; Michaels and Bazzaz, 1986).

Most of the debate on how sexual reproduction may provide an advantage over asexual reproduction has centred on a single issue – that of the advantages to be gained by a sexual individual through genetic recombination (crossing-over). Two issues have been focused on. The first is the role of recombination in repairing DNA or in otherwise restoring DNA to an optimal condition. The second concerns the advantages to be gained from recombination resulting in the generation of adaptive combinations of genes.

The DNA repair hypothesis argues that genetic recombination is primarily selected for the repair of DNA, with damage on one chromosome being repaired with reference to the equivalent undamaged sequence on the homologous chromosome (Dougherty, 1955; Bernstein, 1977, 1983; Bernstein *et al.*, 1981, 1984, 1985a,b; 1988). According to this hypothesis, it is double-strand damage rather than single-strand damage that is the selective force for the maintenance of genetic recombination, as only repair of the former requires the presence of a homologous molecule. The requirement for a homologous molecule provides the selective force for the establishment of diploidy. But repair of double-strand damage typically involves crossing over between outside markers, which generates homozygosity at loci distal to the crossover. A potentially serious deleterious consequence of this is that homozygosity will reduce fitness if it exposes deleterious recessive mutations. Sex will be favoured as a means of counteracting this as it restores heterozygosity. Maynard Smith (1988a) points out two problems with this argument. First, double-strand repair should not necessarily require crossing over between outside markers and consequently the argument that sex has evolved to enable heterozygosity to be restored following repair is unnecessarily complicated. Second, a large number of eukaryotes, including asyndetic male *Drosophila* and many asexual taxa which have mitotized or asyndetic meioses, do not undergo recombination in the germ line and yet produce viable gametes; this indicates either that double-strand damage in the germ line is rare or unimportant or that it can be repaired by a process not involving crossing-over. Because of this, Maynard Smith concludes that the argument that recombination is maintained primarily for the repair of double-strand damage to DNA fails.

Holliday (1984) has argued that genetic recombination may be primarily selected to allow DNA that has been demethylated to become methylated, using the methylated homologous molecule as the template. In contrast, Bengtsson (1985) has suggested that recombination may allow the repair of DNA through gene conversion. However, as Maynard Smith (1988a) points out, these hypotheses suffer from the same weakness as that of the DNA repair hypothesis described above. They can explain the advantage of diploidy but not of recombinational meiosis as this latter phenomenon is absent from many eukaryotes.

Basically, DNA repair hypotheses suffer from an inherent weakness. They argue that recombination is sufficiently important to confer a selective advantage to the process that it is associated with (i.e. sexual reproduction), but they fail to explain why or how, if it is so important, thousands of asexual species are able to persist successfully, in the short term at least, without it. They seem to be saying, in effect, but with no real explanation, that organisms that exhibit recombination have a requirement for it for DNA repair but those that lack it do not. This is not a convincing argument.

The second aspect of recombination that has been focused on in the attempt to identify the advantages of sexual reproduction has been its role in generating progeny with new adaptive combinations of genes. Felsenstein (1988) has played a central role in bringing together and contributing ideas on how recombination may fulfil this role, and the first part of the following discussion draws heavily on this paper. The arguments consider the situation when fitness effects between loci are multiplicative (multiplicative selection). This situation arises whenever selection at one locus occurs at a different part of the life cycle to selection at another. For example, consider a haploid population in which locus A affects embryo survival and locus B juvenile survival. Let genes $A1$ and $A2$ have relative fitnesses of 1 and 0.9 and genes $B1$ and $B2$ have relative fitnesses of 1 and 0.8. Initially, let there be 100 $A1B1$ embryos and the same number of $A2B2$ embryos. Of these, all 100 $A1B1$ embryos but only 90 $A2B2$ embryos will survive to become juveniles. Of these, all 100 $A1B1$ juveniles but only a proportion 0.8 of the 90 $A2B2$ juveniles ($= 72$) will survive to reproductive maturity. Thus the overall relative fitness of $A1B1$ to $A2B2$ is 1:0.72, which can be rewritten as $(1)(1):(0.9)(0.8)$ – the product of the fitnesses of the alleles in each genome.

Felsenstein (1965, 1974, 1988; Felsenstein and Yokoyama, 1976) describes how, with multiplicative selection, the effect of recombination is to reduce linkage disequilibrium (the non-random association between genotypes at different loci) towards zero. In other words, recombination generates linkage equilibrium (the random association between genotypes at different loci). It has this effect by creating gametes that contain genes randomly assembled from different gametes of the previous generation. Of course, genes at loci on non-homologous chromosomes are associated randomly in genomes, as non-homologous chromosomes segregate independently at meiosis. Thus, the adaptive role being sought here for recombination concerns its capacity to influence the association between genes at different loci on the same chromosome. For example, consider that loci A and B, with alleles A_1, A_2 and B_1, B_2, are on the same chromosome. Assume that, initially, a chromosome can be either A_1B_1 or A_2B_2. Only three diploid genotypes are possible: $A_1A_1B_1B_1$, $A_1A_2B_1B_2$ and $A_2A_2B_2B_2$. If there is no recombination between these loci only A_1B_1 and A_2B_2 gametes will be generated and the progeny generation will exhibit the same three diploid genotypes as the parent generation. But if a cross-over (chiasmata) occurs between these loci the new gamete types A_1B_2 and A_2B_1 will be generated. These will be able to fuse with A_1B_1 and A_2B_2 gametes to give four new diploid genotypes: $A_1A_1B_1B_2$, $A_1A_2B_1B_1$, $A_2A_2B_1B_2$ and $A_1A_2B_2B_2$.

The importance of this is that the strength of selection is influenced by the extent to which loci are in linkage equilibrium or disequilibrium. With linkage equilibrium, selection acting to change the gene or genotype frequency at one locus will not cause the gene or genotype frequency to be changed at another locus. In contrast, with linkage disequilibrium, selection at one locus will affect the gene or genotype frequency at a linked locus.

This can be seen if the two loci (A and B) are considered again. First, consider that the loci are in linkage equilibrium. Let the initial frequency (i.e. the frequency before selection) of these alleles be $p(A_1)$, $p(A_2)$, $p(B_1)$ and $p(B_2)$, and let the frequency of the genotypes at each locus be $p(A_1A_1)$, $p(A_1A_2)$, $p(A_2A_2)$ and $p(B_1B_1)$, $p(B_1B_2)$, $p(B_2B_2)$. Then, because the alleles at one locus segregate independently with respect to the alleles at the other, the frequency of $A_1A_2B_1B_1$ is simply the product of the frequencies of the genotypes at each locus – $p(A_1A_2)p(B_1B_1)$. Similarly, the frequency of $A_1A_2B_1B_2$ is $p(A_1A_2)p(B_1B_2)$, that of $A_1A_2B_2B_2$ is $p(A_1A_2)p(B_2B_2)$, and so on. If we use the following arbitrary values for the frequencies of the different genes

$$p(A_1) = 0.2$$
$$p(A_2) = 1 - p(A_1) = 0.8$$
$$p(B_1) = 0.6$$
$$p(B_2) = 1 - p(B_1) = 0.4$$

then the following genotype frequencies per locus are given (from Hardy-Weinberg)

$$p(A_1A_1) = 0.2^2 = 0.04$$
$$p(A_1A_2) = 2(0.2)(0.8) = 0.32$$
$$p(A_2A_2) = 0.8^2 = 0.64$$

$p(B_1B_1) = 0.6^2 = 0.36$
$p(B_1B_2) = 2(0.6)(0.4) = 0.48$
$p(B_2B_2) = 0.4^2 = 0.16$

With linkage equilibrium, the genotypes at the two loci come together at their random frequencies. Thus

$A_1A_1B_1B_1 = (0.04)(0.36)$
$A_1A_1B_1B_2 = (0.04)(0.48)$
$A_1A_1B_2B_2 = (0.04)(0.16)$
$A_1A_2B_1B_1 = (0.32)(0.36)$
$A_1A_2B_1B_2 = (0.32)(0.48)$
$A_1A_2B_2B_2 = (0.32)(0.16)$
$A_2A_2B_1B_1 = (0.64)(0.36)$
$A_2A_2B_1B_2 = (0.64)(0.48)$
$A_2A_2B_2B_2 = (0.64)(0.16)$

It can be seen that the proportion of B_1 and B_2 alleles in A_2A_2 genomes is the same as their proportion in A_1A_2 and A_1A_1 genomes: B_1 is present twice in 36% and once in 48% of A_1A_1 genomes *and* of A_1A_2 genomes *and* of A_2A_2 genomes. Thus selection at the A locus will not affect the population frequency of the two alleles at the B locus: the frequency of B_1 and B_2 alleles will not be affected by whether, for example, A_1A_1 is better or worse than A_1A_2 or A_2A_2. The situation would be very different if the two loci were in linkage disequilibrium. Consider an extreme example of disequilibrium in which all individuals in the population are either $A_1A_1B_2B_2$ or $A_2A_2B_1B_1$. Here, selection against the A_2 allele will automatically reduce the frequency of the B_1 allele in the population relative to that of the B_2 allele. Conversely, selection in favour of the A_2 allele will automatically increase the frequency of the B_1 allele.

It is clear from this that the rate of response of a locus to selection is affected by whether or not there is linkage disequilibrium. In its absence, the rate of change in the frequency of an allele is determined solely by the selective relationship between it and other alleles at the same locus. In its presence, it is also affected by the effects of selection on associated loci.

The overall effect of linkage disequilibrium is profound. It alters the rate of response of a population to selection. The genetic equilibria in a population will be affected, and the rate of response to selection slowed, if linkage disequilibrium causes an association between a selectively deleterious allele at one locus and a selectively favourable allele at another. This is an example of 'repulsion' linkage disequilibrium. In contrast, if selectively advantageous alleles at different loci are associated in 'coupling' linkage disequilibrium the rate of response of a population to selection will be increased.

Thus, recombination reduces linkage disequilibrium, but the rate of response of a population to selection, and the genetic equilibria in the population, are affected by whether or not there is linkage disequilibrium. It follows from this that, with multiplicative selection, if recombination is to have any continuous effect on the rate of response of a population to selection (i.e. on the rate of evolution) or on the genetic equilibria of populations linkage disequilibrium must be being continuously generated somehow. Otherwise, recombination would eventually generate linkage equilibrium, at which time it would cease to have any effect on the genetic equilibria in populations or on the rate of evolution. Moreover, for recombination to be favoured (i.e. maintained by selection) linkage disequilibrium must be maladaptive. Otherwise, the effect of recombination in reducing it will be deleterious, and recombination will be selected against.

One way in which linkage disequilibrium can be continuously generated in a finite population is by genetic drift (Felsenstein, 1985). Disequilibrium generated in this way will cause favoured alleles at different loci to be sometimes in coupling and sometimes in repulsion disequilibrium. Clearly, for recombination to be favoured for its effects on reducing disequilibrium, the benefits obtained from coupling disequilibrium must be outweighed by the costs resulting from repulsion disequilibrium. Hill and Robertson (1966) demonstrate that this is the case and show that the overall effect of disequilibrium is a reduction in the rate of response of a population to selection. That is, the accelerating effects due to coupling linkage disequilibrium on the rate of response to selection can only partly compensate for the slowing of this rate caused by repulsion linkage disequilibrium. Recombination, by generating linkage equilibrium, will therefore generate a higher rate of response to selection at these loci.

Another way in which linkage disequilibrium can be continuously generated is by natural selection. Sturtevant and Mather (1938, discussed in Felsenstein, 1988) consider the situation in which natural selection favours gametes A_1B_1 and A_2B_2 in some generations and gametes A_2B_1 and A_1B_2 in others. If selection favours one combination for a long enough period the population would become fixed for it. In this situation, the only way of generating the other combination when it became selectively advantageous would be by recombination.

Finally, Maynard Smith (1980) has demonstrated that normalizing selection for a quantitative trait will lead to linkage disequilibrium. If the selective optimum remains constant, homozygosity will eventually result, but if the optimum changes, either in a fluctuating fashion or in a constant direction, recombination will be favourably selected. In a further paper, Maynard Smith (1988b) looks more closely at the effects on the rate of recombination of directional and normalizing selection and of selection with a fluctuating optimum on a quantitative trait. He finds that, for plausible selection schemes, directional selection generates repulsion linkage disequilibrium, but that this is less strong in gametes carrying a gene for high recombination than in gametes carrying a gene for low recombination (because recombination destroys disequilibrium). Consequently, the mean and the phenotypic variance of high recombination chromosomes is higher than that of low recombination ones and, because of their higher variance, high recombination chromosomes respond more to selection, accumulating selectively favoured alleles. Genes for high recombination then increase in frequency by hitch-hiking with the selectively favoured alleles they have attracted. The spread of genes for high and low recombination is more difficult to predict when the direction of selection fluctuates. Overall, high recombination seems to be favoured if changes in the direction of selection are infrequent. Genes for low recombination are favoured if changes in the direction of selection are frequent, or if there is normalizing selection. Empirical support for this model is provided by Burt and Bell (1987) who show that excess chiasma number (i.e. chiasma number minus haploid chromosome number) in male mammals is highest in domesticated breeds, which clearly have been subjected to many generations of directional selection.

The account given above describes in rather general terms the ways in which patterns of interactions between genes at different loci may be generated and may experience favourable or unfavourable selection. However, a number of models have been proposed that weld these genetic arguments to an ecological framework. Using Bell's (1982) terminology, these models fall into two categories. They are Tangled Bank models, or Red Queen models. Tangled Bank models argue that recombination is favoured because the environment a population occupies varies widely between patches, so much so that different genotypes are optimal in different patches. Red Queen models, which include Maynard Smith's (1980, 1988a,b) Shifting Optimum model, argue that recombination is favoured because the environment changes in time (rather than in space), and in such a way that a genotype that is advantageous in one generation is deleterious in another. Interactions between hosts and their antagonists (e.g. parasites, predators or pathogens) may easily generate this situation (Levin, 1975; Jaenike, 1978; Hamilton, 1980, 1982; Rice, 1983; Bremermann, 1980; Hamilton *et al.*, 1990).

It is clear that an environment that is sufficiently variable in space and/or in time provides conditions under which recombination (through crossing-over) would be adaptive. But it is not yet clear whether most environments are sufficiently variable to account for the ubiquity of genetic recombination. Indeed, empirical ecological investigations have so far largely failed to demonstrate any consistent ecological advantage of sexuality (and thus of genetic recombination) over asexuality. For example, Antonovics and co-workers have shown, for the sexual, clonally growing grass *Anthoxanthum odoratum*, that uncommon (minority) genotypes when grown amid common (majority) genotypes can be much fitter than the latter and that sexually produced offspring can be fitter than ramets produced by clonal growth (Antonovics and Ellstrand, 1984; Ellstrand and Antonovics, 1985; Schmitt and Antonovics, 1986). However, the advantages accruing to minority genotypes were obtained irrespective of whether these genotypes were produced by clonal growth or by sexual reproduction: rarity appeared to be a more important determinant of fitness than means of production. And Bierzychudek (1987a) has pointed out that the advantages exhibited by sexually produced offspring over ramets may be due to the sexually produced offspring having been produced by seed, during which stage they may have been able to shed any pathogens carried by their mother. The vegetatively propagated progeny would not have had access to this cleansing system. If this is the case, then asexual progeny produced from seed would have experienced the same advantage over ramets as the sexually produced progeny. Bierzychudek (1987a: 172) concludes from an appraisal of these and other investigations into the ecological advantages of sex that

> So far, none of the experimental work that has been conducted has produced particularly satisfying results. Some experiments have been unable to demonstrate that sexual progeny really do enjoy an advantage over asexual progeny; others have succeeded in making this demonstration, but have not led to an understanding of a mechanism of the observed advantage.

However, problems would remain even if the empirical data were more supportive of the models. Thus Red Queen and Tangled Bank models propose that genetic recombination is favourably selected because of its effects on progeny. It provides these with genotypes that will help them to survive the demands of an unforgiving and variable and varying environment. But while there is no doubt that recombination, by reducing linkage disequilibrium, can have this effect it becomes difficult to understand how, if this is the primary reason for its maintenance, asexual taxa can co-exist, as they do, with sexual taxa. If the latter need genetic recombination to cope with the vagaries of the environment then why not the former? Nor is it clear why, in the great majority of sexual taxa, genetic recombination should be (as it seems to be) characteristic of every reductional meiosis of every individual of every population. There are exceptions among sexual taxa to this rule although these are relatively few. They include some oligochaete worms which are characterized by achiasmate meioses in both sexes (Christensen, 1961), and a number of *Diptera*, grasshoppers, bugs, butterflies and moths and the beetle *Caraboidea* and the mantid *Callimantis antillarum*, where the heterogametic sex is achiasmate but the homogametic sex is chiasmate (White, 1938, 1965; Serrano, 1981; Nokkala and Nokkala 1983, 1984). This phenomenon is discussed by Bell (1982) and Trivers (1988). But within a taxon no sex or individual appears to practise a mixed strategy of chiasmate and achiasmate reductional meiosis. The absence of such a mixed strategy from individuals is most unexpected, given that facultative apomixis and a mixed strategy of outcrossing and selfing are genetically analogous in many respects to the absent chiasmate/achiasmate strategy and are clearly successful.

The absence of achiasmate reductional meiosis from the overwhelming majority of sexual taxa and the absence of a mixed strategy of chiasmate/achiasmate meiosis in individuals are unlikely to be due to there being a general lack of capacity in sexual populations or within individuals to vary the rate of genetic recombination. Indeed, as Maynard Smith (1978) observes, experiments which attempt to alter recombination frequency by selection usually succeed (e.g. Allard, 1963; Dewees, 1970; Gale and Rees, 1970; Shaw, 1972; Abdullah and Charlesworth, 1974; Charlesworth and Charlesworth 1985a,b; and review by Brooks, 1988), and Bell (1982: 413–424) and Trivers (1988) describe how differences in chiasma frequency and location are common, being found between males and females of many dioecious taxa, between male and female meiosis of some cosexual taxa (e.g. Ved Brat, 1966; Moran *et al.*, 1983), and between different populations of many taxa. But apart from the exceptions mentioned above, this variation does not encompass an absence of chiasma.

The conclusion to be drawn from this is that genetic recombination is ubiquitous not because it is unavoidable but because it is maintained by selection. But what is the nature of the selection pressure maintaining it? I find it difficult to accept that the primary selection pressure is usually the environment of the offspring, as argued by Red Queen and Tangled Bank models. If it is the offspring's environment how can asexual taxa coexist with sexual taxa in a great variety of habitats? And how can asexual taxa persist for generation after generation in habitats that had been occupied by their sexual ancestors, who presumably underwent chiasmate meioses, and often in habitats that are still occupied by sexual conspecifics or congenerics who still do? And how can the asexually produced offspring of facultative apom-

icts coexist with their sexually produced sibs? Basically, the hypothesis that the selection pressure to maintain genetic recombination
is imposed by an offspring's environment has weak foundations. It
may explain why asexual taxa are absent from, or rare in, some habitats that support large and diverse sexual communities. But it does not
explain why genetic recombination is ubiquitous among the sexual
component of communities that contain, indeed may even be dominated ecologically by, a large asexual component.

6.3   GENETIC RECOMBINATION AND MEIOTIC REDUCTION

Genetic recombination is associated with the exchange of flanking
markers and consequently with the generation of new gene sequences.
Thus a cross-over between loci B and C involving chromatids with
the gene sequences $A_1B_1C_1D_1$ and $A_2B_2C_2D_2$ will generate the new
gene sequences $A_1B_1C_2D_2$ and $A_2B_2C_1D_1$. This is shown in generalized form in Fig. 6.1(a–c). Given that this association need not occur
it is reasonable to assume that it is maintained by natural selection.
It was shown in section 6.2 how this consequence of recombination
can benefit offspring if, for example, the linkage of A_1 with C_1 and
A_2 with C_2 is adaptive in one patch or generation but the linkage of
A_1 with C_2 and A_2 with C_1 is adaptive in another. But doubt has been
cast on this phenomenon being generally capable of accounting for
the maintenance of genetic recombination. However, this does not
give sufficient cause to dismiss the hypothesis that the advantage of
genetic recombination is derived from the exchange of flanking markers. It simply gives cause to doubt that the prime beneficiary of these
exchanges has been correctly identified.

The scope for misidentifying the prime beneficiary can be gauged
by considering what, for want of a better term, I will describe as the
'life history' of a gene sequence that has been newly generated by
recombination. It is generated during prophase I of the parental meiosis, where it forms part of a chromatid of one of a pair of homologous
chromosomes. For example, R_1 and R_2 in Fig. 6.1 are chromatids that
contain new gene sequences obtained, through crossing-over, from the
parental sequences A_1 and A_2. Thus the first environment of a newly
generated gene sequence is a meiotically dividing but still diploid cell.
As meiosis progresses, first the homologous chromosomes and then
the chromatids of each chromosome separate and are included in
different nuclei. Thus the gene sequence moves in stages from a diploid nuclear environment to a haploid one. It will remain in the
haploid environment for a period that is lengthy in some taxa (e.g. the
long-lived gametophyte stage of bryophytes and gymnosperms) but
short in others (e.g. the gametophyte stage of flowering plants and the

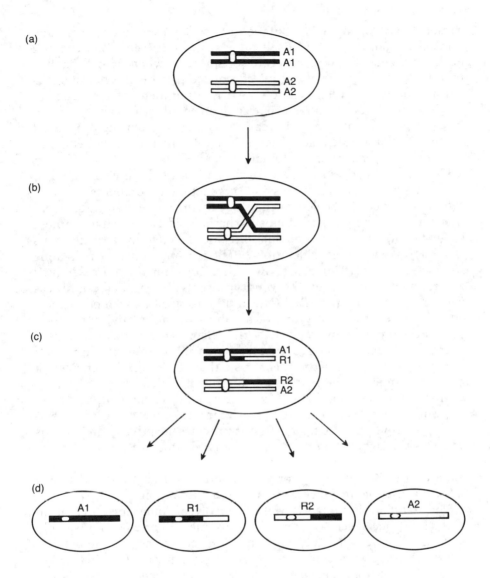

Figure 6.1 A generalized diagram of the effects of crossing-over during meiosis. A single bivalent is considered in which one of the chromosomes carries gene sequence A1 and its homologue gene sequence A2 (a). A single crossover occurs (b) to produce, at the dyad chromosome stage, recombinant sequences R1 and R2 together with the original sequences A1 and A2 (c). The four meiotic products each receive a single gene sequence (d). Note from (c) that heterozygosity is maintained within chromosomes during the dyad chromosome stage at points distal to the cross-over point, but is maintained between chromosomes at points proximal to the cross-over point.

gamete stage of animals). But sooner or later, following cell and nuclear fusion, it will find itself once again in the environment of a diploid nucleus – that of a zygote (offspring) – where it will be able to interact with its newly acquired counterpart on the homologous chromosome. Eventually, following the attainment of reproductive maturity by the offspring, the gene sequence will once again find itself in the environment of a meiotically dividing cell. During prophase I of this division its life history will effectively end if the gene sequence is disrupted by genetic recombination.

Clearly, for a gene sequence to be successful it must be able to perform adequately in a succession of genetic environments. That is, a gene sequence can be described as occupying a fine-grained environment: that of the parental meiosis, the haploid gametophyte/gamete stage, the diploid stage of the offspring and the offspring meiosis. This is rarely taken into consideration but it is important that it is done so, as Strobeck (1975) has shown that selection in a fine-grained environment will favour the genotype that has the highest fitness over the whole lifespan. Red Queen and Tangled Bank models have effectively identified only one of a gene sequence's environments – that of the diploid growth phase of the offspring – and have sought an adaptive explanation for genetic recombination within its confines. It is because of this that there is a risk that the prime beneficiary of genetic recombination may have been misidentified.

In the next few pages I will look at genetic recombination afresh. In doing so I will argue that the primary adaptive value of genetic recombination in many taxa is that it provides adaptive combinations of genes throughout most of a reductional meiosis. That is, it is the environment of a meiotically dividing cell, rather than the environment of the haploid phase or the offspring's environment that can select for the maintenance of genetic recombination.

The argument is straightforward. It revolves around the observation that a sequence of genes along a chromosome will often comprise only half an adaptive sequence, the other half being the complementary sequence along the homologue of this chromosome. This is because homologous chromosomes may carry different genes, and it is the interactions between these that are beneficial. These interactions will largely involve allozygous effects such as genetic complementation and heterozygous superiority. But genetic recombination must occur if these effects are to be retained by the two genomes formed at anaphase I of meiosis, for in its absence each dyad stage nucleus will receive two copies of one of the genes that were present at an allozygous locus in the generative cell rather than one copy of each of the genes present at this locus. This can be seen in Fig. 6.1 where allozygosity is distributed between homologous chromosomes in those parts of the chromosomes that are proximal to the cross-over point and thus that are not affected by genetic recombination but is maintained within each of the pair of chromosomes (being distributed between the chromatids of a chromosome) in those parts that are distal to the cross-over point and that are thus affected by genetic recombination. These allozygous effects will be retained until chromatid separation at anaphase II of meiosis, at which time the chromatids of a chromosome are passed into different nuclei, where they adopt the role of chromosomes. I will refer to the period between anaphase I and anaphase II as the dyad chromosome stage.

In effect, I am proposing that the primary adaptive value of genetic recombination in many taxa is that it allows more than half the different types of genes present in the diploid genome to be retained by each of the nuclei produced by the first division of meiosis, even though each of these receives only a haploid complement of chromosomes. Genetic recombination provides the means by which these nuclei can effectively hang on to any benefits of diploidy.

A great strength of this hypothesis is that it explains why genetic recombination should be a characteristic of every reductional meiosis. It can be seen from Fig. 6.1(c) that the new gene sequences on chromatids R_1 and R_2 are adaptive because they cause allozygosity to become a within-chromosome phenomenon. But this effect is lost following chromatid separation (Fig. 6.1d). For simplicity, only the haploid genome in Fig. 6.1(d) that contains chromosome R_1 will be considered here. A gamete containing this chromosome will fuse with another gamete from the population to form a diploid zygote. Assume that the chromosome in this other gamete that is homologous to R_1 contains gene sequence X_1, so that the new diploid individual (the offspring) can be described as R_1R_1/X_1X_1 (taking into account the presence of two chromatids per chromosome). It can be seen from a comparison of Figs 6.1 and 6.2 that although the gene sequence on R_1 was adaptive during the parental meiosis it will be maladaptive during the offspring meiosis. It was adaptive during the parental meiosis because it contributed to the maintenance of allozygosity during the first division of meiosis. It will have the opposite effect during meiosis in the offspring. This change in its value and in its effect reflects the change that has occurred in its status between the parental and offspring meioses. In the former it was a post-recombinational sequence, in the latter it is a pre-recombinational sequence. If the benefits of allozygosity are to be retained by the dyad chromosome stage of the offspring meiosis, R_1 will have to be disrupted by crossing-over, else one dyad nucleus will receive the non-recombinant autozygous sequence X_1/X_1 and the other the non-recombinant autozygous sequence R_1/R_1. Thus, if there is a selective advantage to retaining allozygosity at the dyad chromo-

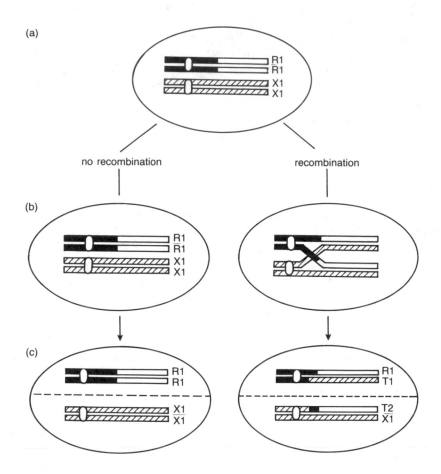

Figure 6.2 This figure follows the fate of gene sequence R1, that was generated as a recombinant sequence in Fig. 6.1. A gamete containing R1 fused
with a gamete containing homologous gene sequence X1 to give the diploid
cell depicted in part (a). If this cell undergoes meiosis in the absence of
genetic recombination heterozygosity is maintained between, but not within,
chromosomes at the dyad chromosome stage (c). In the presence of recombination heterozygosity is maintained within chromosomes at points distal to
the cross-over point (c). Thus the gene sequence R1, which was instrumental
in maintaining within-chromosome heterozygosity during the dyad chromosome stage in the parental meiosis (Fig. 6.1) will not fulfil this role during the
offspring meiosis. R1 must be recombined to generate new sequences T1 and
T2 if within-chromosome heterozygosity is to be maintained during the dyad
chromosome stage of the offspring meiosis (c).

some stage, this will be translated into a selective advantage to retain genetic recombination through crossing-over in every reductional meiosis each generation, providing a generation-by-generation, meiosis-by-meiosis, advantage to genetic recombination.

This hypothesis is a courtier of the Red Queen, as it proposes a switch in the selective value of a gene combination between generations that will result in the maintenance of genetic recombination. It differs from other attendant hypotheses in the timing of the switch and in the nature of the environment imposing the selection pressure. Rather than this pressure being imposed postzygotically, by changes between generations in the physical or biotic environment, it is first imposed prezygotically, indeed pregametically, by changes between generations in the nuclear environment of the meiotically dividing generative cell. An interesting consequence of this shift in timing is that the beneficial role that genetic recombination plays during meiosis is the opposite to the role it is credited with in models that propose that its adaptive value is that it improves the ecological performance of offspring. It fulfils the latter role by generating non-parental combinations of genes. But it fulfils its adaptive role during meiosis by maintaining parental combinations of genes. Note that a gene for recombination that has been favourably selected because it generates a healthy, allozygous genetic environment at the dyad chromosome stage generates this environment in each of the sister genomes initiated at anaphase I, as each of these receives the same fitness-conferring allozygous sequence (Fig. 6.1c). The gene cannot, therefore, be separated from its beneficial effects. This is the case irrespective of whether it induces recombination within its own bivalent or within others. This offers an adaptive explanation for why many genes for recombination induce it over all or much of the genome, rather than simply on the chromosomes on which they reside.

The model has its attractions but it is also associated with some problems. One relates to the fitness effects of recombination on post-meiotic stages. The validity of the model is affected by whether it is reasonable to assume that the advantages of recombination at the dyad chromosome stage are sufficient to outweigh any disadvantages that recombination might generate during the stages that separate meiosis in the parent from meiosis in the progeny. In a very real sense, there is a need here to introduce the Tangled Bank as the consort of the Red Queen. Recombination must prove advantageous to each generation, by helping each to optimize meiosis, but it must also prove advantageous during each generation, as the recombined genomes traverse the patchy environment of the different stages of the life cycle. I believe that recombination can meet these diverse requirements. In most environments, selection to maintain a gene sequence may be strong only during the haploid phase, as the adaptive value of a sequence during the parental meiosis and during the offspring growth stage will be determined in part by how well it interacts with its partner sequence on the homologous chromosome. With outcrossing, the probability of these same partner sequences coming together again in the offspring generation will typically be extremely low, even if there is no crossing-over. The disruption of adaptive partner sequences is much more a cost of meiotic reduction than of crossing-over. Consequently, any advantages accruing to the post-meiotic stages by suppressing crossing-over could be easily outweighed by the advantages accruing to the meiotic stage through recombination.

A more formidable problem is that the proposed selective advantage to recombination requires that there is gene activity at the dyad chromosome stage. If this assumption is incorrect the model is fatally weakened. Even if gene activity can be demonstrated, there is one further problem that needs to be resolved. This concerns the capacity of the nuclei produced during a meiotic division to function together as a single transcriptional unit (i.e. to function as though they were still members of the same diploid nucleus). In many plants and some animals, each of the nuclei produced by the first division of meiosis is located in a different cell. It seems reasonable to assume that these will not be able to interact as a single transcriptional unit as messenger RNAs are unlikely to cross cell membranes or callose-coated cell walls; consequently, each of the cells will benefit from allozygosity only if crossing-over has occurred. However, in some plants and many animals the dyad nuclei remain in a common cytoplasm. For each of these to obtain benefit from crossing-over at the dyad chromosome stage, maintaining allozygosity within a nucleus must be more advantageous than maintaining it between nuclei. This requires that the two nuclei do not act as a single transcriptional unit. Unfortunately, there is a general lack of information about either of these problems. But their consideration can proceed by inference. I will deal with the second problem first.

In the absence of any direct information, the problem of whether dyad nuclei that occupy the same cytoplasm act independently of each other or as a single transcriptional unit must be approached laterally, by gleaning information from analogous systems about which there is more information. The most promising approach is to consider the consequences of gene activity in heterokaryotic, dikaryotic and diploid fungal cells. The dikaryotic cell contains two haploid nuclei which are often genetically different, having been brought together following fusion between different mycelia. The heterokaryotic cell contains several haploid nuclei. Thus each locus of a nucleus of a dikaryon or heterokaryon is represented by a single allele which is present on each chromatid of the chromosome. Consequently, the dikaryotic or heterokaryotic nucleus is genetically equivalent to a dyad chromosome stage genome that has not been recombined. In both, allozygosity is exhibited between but not within nuclei. If the dikaryon, heterokaryon or unrecombined meiotic dyad stage is to benefit from allozygosity (e.g. as a result of complementation or heterosis), then neighbouring nuclei must act as a single transcriptional unit.

A number of investigations have been carried out to determine whether, and to what extent, this is the case in dikaryons and heterokaryons. Detailed discussions provided by Burnett (1975) and Fincham et al. (1979) show that, on the whole, interactions are typically of a type expected in a diploid. That is, the nuclei behave as a single transcriptional unit. However, there are exceptions. For example, Apirion (1966) found that complementation occurred in diploid but not in heterokaryotic cells formed following the fusion of auxotrophic mutants of Aspergillus nidulans. And Casselton and Lewis (1967) found similar levels of complementation in dikaryons and diploids but lower levels in heterokaryons of Coprinus lagopus. A subsequent study found that the two genomes associated in a dikaryotic cell of Coprinus may produce different products to those produced when these genomes are associated in a diploid nucleus (Senathirajah and Lewis, 1975).

Gene expression in a dikaryon or heterokaryon can differ from that in a diploid either because certain products of gene interaction have to be made in the nucleus rather than in the cytoplasm, or because the products of each nucleus do not become sufficiently intermingled (Fincham et al., 1979). The second reason explains why it is more likely that dikaryotic rather than heterokaryotic associations should behave like diploid associations. The two nuclei in a dikaryotic cell are very closely associated and divide synchronously. In contrast, the nuclei in a heterokaryotic cell are not closely associated and often divide asynchronously. This is a crucial observation, as the two nuclei of the meiotic dyad stage are typically not closely associated. Moreover, they often do not divide in synchrony. Indeed, during meiosis in female animals and in a few flowering plants, only one of the dyad nuclei normally proceeds with the second meiotic division (Bouman, 1984); in animals the other degenerates into the first polar body. Thus, studies of heterokaryotic fungi provide some strong circumstantial evidence that the arrangement of, and lack of synchrony between, the nuclei of a meiotically dividing cell during the dyad chromosome stage may be sufficient to preclude them acting as a single transcriptional unit when they are occupying the same cytoplasm.

There are two other reasons for doubting that dyad chromosome stage nuclei interact with each other when they are occupying the same cytoplasm. The first is that meiotically dividing cells are often highly polarized and the cytoplasm is often allocated in different proportions to each dyad nucleus (Willemse and van Went, 1984). The nuclei are therefore in different environments and these may induce different levels of gene activity. The polarization may also preclude the intermingling of their products. The second is that, although the two nuclei are included within the same cell membrane, it is likely that they do not effectively share the same cytoplasm. A furrowing of the cytoplasm between the two nuclei has been observed on numerous occasions (Bhandari, 1984). This furrowing may be as effective as a cell wall at isolating the nuclei and their products from each other.

Thus there is quite strong circumstantial evidence that the events associated with the dyad chromosome stage of meiosis are such that the two nuclei will not act as a single transcriptional unit. Although they may be physically located within the same cell, they may not be equally active, and they may not effectively share the same cytoplasm. If they are to retain the benefits of allozygosity, it will be necessary for each of them to retain this condition at their active loci. This requires that these loci are genetically recombined.

This leaves as the major challenge to the validity of the model the problem of whether gene activity during the dyad chromosome stage is sufficiently common to provide a selective advantage to crossing-over. There is a paucity of information on this, although the observation that this stage in many organisms includes an interphase (interkinesis), during which the coiling of the DNA is relaxed as though preparation is being made for gene expression, is strongly indicative of this. Unfortunately, it is not clear whether genes are expressed during interkinesis which, in any case, is not exhibited by many organisms.

Certainly, the dyad chromosome stage is bounded by stages that exhibit considerable gene activity. The chromosomes of many organisms adopt the 'lamp-brush' configuration during prophase of the first division. This configuration is associated with intense transcriptional activity. In animals, many of the RNAs produced during this period remain untranslated until the gametic or embryonic stages (Davidson, 1976; Cohen, 1977; Bostock and Sumner, 1978). This pattern of transcription and translation may be indicative that the inexorable drift towards full haploid status during meiosis is a weak point best left unexposed. However, an alternative explanation is that the organism is simply provisioning its gametes or embryos with expensive, well-tested products in order to confer on them advantages over competitors. Whereas the former view leads to the conclusion that genetic activity should be generally suppressed following chromosome separation at meiosis, and not reinstated until after zygote formation, the latter does not. The former view is not tenable in general. The postmeiotic haploid phase in many organisms is prolonged and sometimes complex, and must be characterized by much gene activity. The gametophytic stage in plants, and the haploid males of social insects are obvious examples. However, evidence for gene expression during the postmeiotic phase in animals with a simple unicellular (gametic) haploid phase is sparse. Nevertheless, there are a few examples of gene expression in spermatids and spermatozoa (Hogarth, 1978; Stewart *et al.*, 1988) and there is some evidence that some aspects of sperm competition may be directed by the sperm's genome (Beatty, 1975; and references in Smith, 1984). Most examples of meiotic drive (e.g. Beatty, 1975; Turner and Perkins, 1979; Swanson *et al.*, 1981) also appear to require gene expression either at the dyad chromosome stage or during the haploid phase following meiosis, as does Cohen's (1977) hypothesis that gametic redundancy largely results from the weeding out of meiotic products with inviable genomes.

But evidence that genes are active during the haploid gametic or gametophytic phase does not provide firm grounds for assuming that they are also active during the dyad chromosome stage of meiosis. This stage differs from the haploid stage in a way that could favour the expression of genes in the latter but their suppression in the former. Circumstantial evidence that this has not occurred comes from considering these differences and the ways in which organisms have responded to them.

Interactions between meiotic products will, at some stage in most organisms, be between the products of several of an individual's meioses. For example, sperm from different meioses may interact within sperm cysts or within ejaculates, pollen (male gametophytes) from different meioses may interact within pollinia, or males in social insects (derived parthenogenetically from eggs produced by different meioses) may interact within a colony. Products of the different meioses of an individual are closely related, exhibiting a minimum average coefficient of relatedness (in the absence of inbreeding) of 0.5. The interactions should consequently, on the whole, be equitable. However, interactions during meiosis will typically be between products of the same meiosis rather than between products of sister meioses. This is because dyad or tetrad nuclei of the same meiosis are much more closely associated with each other (often sharing the same cell) than with their counterparts in other meiotically dividing cells. The relatedness between the products of a single meiosis can vary tremendously, depending on the pattern and frequency of genetic recombination. In the extreme case of a genome that is allozygous at every locus, sister dyad nuclei may share no genes in common (no genetic recombination, all alleles segregating at meiosis I) or every gene in common (all loci affected by genetic recombination, all alleles segregating at meiosis II). Thus the coefficient of relatedness can range, in theory, from 0 to 1. The same range is obtained under the same conditions among second division products, although the average relatedness between these (though not the variance associated with this mean) will be the same as that between second division products from different meioses. To put it another way, the nuclei produced by a meiosis may be as unrelated as two individuals of phylogenetically distant species, or as closely related as monozygotic twins! Overall, unless each allozygous locus has a 50% probability of being genetically recombined at each meiosis, sister dyad nuclei will be less closely related to each other than they are to their counterparts in other meiotically dividing cells. This can be seen more clearly in Table 6.1, which lists the coefficients of kinship of products of a single meiosis and of sister meioses in the presence and absence of inbreeding and of recombination. It is clear that the potential for conflict between sister dyad nuclei is considerable. Similar conditions favouring conflict can also arise at the tetrad stage, as described by Haigh (1986), and illustrated by Halac and Harte (1975) in *Oenothera*. Indeed, because

Table 6.1   The coefficients of kinship (R) between products of a meiosis and between these and the products of other meioses of the same diploid individual. F is the coefficient of inbreeding. P is the proportion of loci that, through recombination, bear allozygous genes at homologous points on the two chromatids comprising each chromosome of the dyad. P gives the proportion of allozygous loci which segregate at meiosis II rather than at meiosis I. Intrameiotic sister products are tetrad cells derived from the same dyad cell. Intrameiotic non-sister products are tetrad cells derived from different members of the same dyad

Interactants	$R \ (P>0)$	$R \ (P=0)$	Effect of P on R	Equation number
Dyad cell to itself	$1-0.5P$	1	$-0.5P$	(6.1)
Cells within a dyad	$F+0.5P$	F	$+0.5P$	(6.2)
Dyad cell to cell of another dyad	$0.5(1+F)$	$0.5(1+F)$	0	(6.3)
Tetrad cell to itself	1	1	0	(6.4)
Intrameiotic sister tetrad cells	$1-2P(1-P)(1-F)$	1	$-2P(1-P)(1-F)$	(6.5)
Intrameiotic non-sister tetrad cells	$F+P(1-P)(1-F)$	F	$+P(1-P)(1-F)$	(6.6)
Intermeiotic tetrad cells	$0.5(1+F)$	$0.5(1+F)$	0	(6.7)

the products of a meiosis are associated in a highly predictable way, conditions may become suitable for the evolution of spiteful as well as of selfish behaviour.

Spiteful behaviour is characterized by an individual harming itself in order to harm another (conspecific) individual more (Hamilton, 1970; Parker, 1979; Grafen, 1985). Spite may evolve if it is expressed by an actor only when it is interacting with recipients that are related to it by less than the average relatedness of the actor to random members of the population. The actor values the recipient's offspring negatively and will be prepared to forgo some of its own reproduction in order to reduce by a greater amount that of the recipient. Hamilton (1970) has pointed out that examples of spite are rare. One reason is that an organism will not normally be able to recognize which other organisms it is related to by less than average. Another is that a population in which the average relatedness between individuals is much different from zero must typically have a small effective population size and must consequently be in a precarious position that will be made worse by the selection of a gene causing spiteful behaviour. However, neither of these restrictions necessarily applies to a genetically variable population of meiotic products before they leave the parent (i.e. when they are still in the ovary, anther, testis etc.). Here, the population of meiotic products may be hundreds of millions strong (e.g. sperm within an ejaculate, pollen or meiospores produced by a large plant), but the mean relatedness of these is considerably above zero (Table 6.1). Most importantly, there are stages in their production in which it can be predicted that interactions will typically be between interactants which are related by less than the mean relatedness. These are the dyad stage and, to some extent, the tetrad stage.

The pattern of meiosis ensures that the nuclei of the same dyad are at least transiently associated with each other, and that during this stage members of different dyads are, in most organisms, discrete. Thus, within a population of dyad cells, a cell will be less closely related to the cell it is most closely associated with than it is, on average, to other cells drawn randomly from the population. Indeed, in the absence of recombination, the proportion of non-partners that are no more closely related to it than is its partner is only 2^{-n}, where n is the haploid number of chromosomes. Although this difference in relatedness between partner and non-partner nuclei is reduced with inbreeding or with recombination, it does not disappear until $P = 1 - F$ (obtained by setting Equations (6.2) and (6.3) of Table 6.1 equal), where F is the coefficient of inbreeding ($0 \leqslant F \leqslant 1$) and P is the proportion of loci that, through recombination, bear allozygous genes at homologous points on the two chromatids comprising each chromosome of the dyad ($0 \leqslant P \leqslant 1$); that is, P is the proportion of allozygous loci which, because they have been recombined, segregate at meiosis II rather than at meiosis I. In the absence of inbreeding ($F = 0$), this equivalence will not be met unless every allozygous locus is recombined at meiosis I. It is highly unlikely that this will occur.

The conditions necessary to reduce the likelihood of spiteful behaviour between members of a meiotic tetrad differ from those required at the dyad stage. The mean relatedness of a tetrad product to the other products of the tetrad is

$$R = 3^{-1}(1+2F)$$

This will be lower than that between products of different tetrads when $F < 1$ (Equation (6.7) of Table 6.1). However, simply using the means in this way obscures the fact that a tetrad cell is associated with one sister and two non-sister cells, among which the relatednesses differ. The relatedness between a tetrad cell and its sister cell equals the mean relatedness between it and cells produced during other meioses when $P = 0.5$ (setting Equations (6.5) and (6.7) of Table 6.1 equal). The relatedness between non-sister cells cannot be set equal either to the population mean or to that between sister cells of the same tetrad as the former requires $P(1 - P) = 0.5$ and the latter $P(1 - P) = 1$, neither of which is obtainable as P takes values between 0 and 1.

The potential for spiteful behaviour will be modified in certain situations by external ecological factors. The mean relatedness of an individual's meiotic products to each other will be higher than their relatedness to the meiotic products produced by the remainder of the population of which the individual is a part. This latter value will approach that of the coefficient of inbreeding under random outcrossing. Note from Equations (6.2) and (6.6) of Table 6.1 that the least related of an individual's meiotic products will not be less related to each other than they are, on average, to the meiotic products of other individuals. If recombination has occurred they will be more related. Thus the conditions for spiteful behaviour towards sister dyad or tetrad cells will be absent if the consequences of this action are to allow a gamete from another individual, rather than a gamete from another meiosis of the same individual, to gain fitness at the expense of the spiteful actor. This situation may arise in some breeding strategies, for example when the meiotic products are dispersed into the environment where they must compete with each other. The meiospores of homosporous plants and the pollen of heterosporous plants are dispersed in this way, but it is debatable whether the dispersal pattern will favour competition between meiospores/pollen of the same individual rather than between those of different individuals. In many cases, the meiotic products of an individual will only compete against each other, for example, sperm in monogamous matings, pollen in self-fertilizing plants and eggs in females that are internally fertilized. Consequently, conditions that are suitable for the evolution of spiteful behaviour may be widespread.

Conflict between the products of a meiosis, whether as a result of spiteful behaviour or simply of sibling rivalry, may reduce the fitness of the parent. In such cases the evolution of parental strategies that suppress conflict will be expected. Two strategies can be envisaged. The first involves the parent supplying the meiotically dividing cell with the materials (RNAs, enzymes etc.) required for meiosis so that it can suppress all gene activity during this division. This strategy will be tenable only if the materials required by meiosis can be supplied in sufficient quantities, they are sufficiently durable to survive until required, and they can be brought into and out of the metabolic machinery when necessary. If these requirements cannot be satisfied, or if genetic fine-tuning is required during meiosis, some gene activity must be allowed. This will provide genes that cause intrameiotic conflict with the opportunity to hijack the transcriptional machinery of the cell and find expression. The consequences of this happening can be avoided by the parent adopting the second strategy, which involves altering the pattern of meiosis to reduce the potential for intrameiotic conflict. There is a wealth of circumstantial evidence that the latter strategy is the one that has been most commonly adopted. In flowering plants, the parent deposits callose between its dyad cells and between its tetrad cells, greatly reducing the opportunity for these to wage chemical warfare on each other (Haigh, 1986). A similar limitation is imposed in many other organisms. Thus in animals, one of the dyad nuclei produced during female meiosis is shunted into a polar body, and it is common for several cells undergoing male meiosis to be interconnected by cytoplasmic threads so that they effectively share the same cytoplasm. Finally, as already mentioned, sister dyad nuclei often share the same cytoplasm instead of being directed into different dyad cells. The sharing of cytoplasm is an effective countermeasure to competition, if not to spiteful behaviour, as it will result in a debilitating compound affecting the genome of the gene producing it as well as the genomes it is competing against.

There seems little reason to adopt these patterns other than to restrict the opportunity for intrameiotic conflict. None are essential to the basic process of chromosome and chromatid separation or to the subsequent differentiation of the meiotic products. For example, *Pergularia daemia*, in contrast to most other angiosperms, does not produce callose during male meiosis but is nevertheless able to produce viable pollen (Vijayaraghavan and Shukla, 1977). Interestingly, the four pollen grains that result from a meiosis remain closely associated in tetrads rather than being dispersed individually. It is likely that the benefits of co-operation between these may outweigh those that could result from conflict, negating the requirement for the parent plant to take conflict-avoiding measures, by laying down callose, against its meiotic products. A similar situation is exhibited during female meiosis by tetrasporic plants. Most flowering plants are monosporic; here three of the four tetrad cells degenerate and the survivor divides mitotically to form the gametophyte and, eventually, the egg. The tetrad cells of monosporic plants are isolated from each other by callose. In tetrasporic plants, all four tetrad nuclei contribute to the female gametophyte, although only one will give rise to the egg. It is probable that a female gametophyte will develop successfully only if these nuclei do not act aggressively towards each other at the tetrad stage. It is thus of interest that no callose is deposited around a tetrasporic tetrad (Bouman, 1984).

It seems unreasonable to assume that the genome will be maintained in a state conducive to gene activity if the only genes active are those that cause conflict. It is more likely that the latter are provided with the opportunity for expression because the genome must be held in this state to allow expression of genes essential for the control of meiosis. Thus although there is no firm evidence that controlling genes are active during the dyad chromosome stage, there is circumstantial evidence that they may typically be so.

In conclusion, although the conditions required for the model cannot be shown to exist, their presence can be inferred. This is the best that can be done. Thus it is by no means inconceivable, although it remains debatable, that the primary role of crossing-over in many taxa is to provide the mechanism through which allozygosity is maintained during the dyad chromosome stage at loci involved in the control of reductional meiosis. Unfortunately, this brings us no nearer to understanding why sex is maintained. It simply provides a reason why genetic recombination should typically be associated with sex – this being that sex is typically associated with meiotic reduction. This challenges the widely accepted opinion encompassed by other Red Queen models, and by Tangled Bank models, that sexual reproduction is maintained to allow genetic recombination. However, in reality the difference is a fairly minor one. I am not proposing that genetic recombination cannot have profound consequences either for the survival and fitness of sexually reproduced offspring or for DNA repair. Indeed, in habitats in which asexual taxa are rare or absent its primary role may be to equip (sexually produced) offspring with the gene sequences that they will require if they are to survive the ecological pressures imposed by the habitat. But by proposing that its primary role can be to smooth the course of reductional meiosis, it becomes clear why genetic recombination is associated with sexual reproduction irrespective of the spatial or temporal nature of the external environment, irrespective of the requirement for the repair of DNA and irrespective of the fact that asexual taxa that do not exhibit genetic recombination may be common in, and even dominate, some habitat types. It also explains why the success of sexually produced progeny of facultative apomicts appears to be associated with recombination even though that of their asexually produced progeny is not; only the sexual process involves a reductional meiosis.

6.4   THE EVOLUTIONARY POTENTIAL OF ASEXUAL ORGANISMS

A major handicap of obligately asexual lineages is their inability to acquire new genes other than by mutation. One consequence of this is that obligately asexual taxa appear to be short-lived in comparison to related sexual taxa (Darlington, 1939; Stalker, 1956; Muller, 1964; Mayr, 1970; White, 1973). This is indicated by the sporadic taxonomic distribution of asexuality, in which the closest relatives of asexual taxa are usually sexual. One implication that can be made from this is that asexual lines become extinct before they succeed in giving rise to new lines. The situation with respect to longevity will be somewhat different when there is a potential for gene flow through facultative asexuality or via the male function of cosexual asexual forms. In these situations, a gene for asexuality may survive for much longer than if it had caused obligate asexuality, and the apomictic complex may survive for millions of years. Indeed, Stebbins (1971) argues that several apomictic complexes of flowering plants may have survived from the Tertiary period, including *Rubus* subg. *Eubatus*, *Panicum*, *Setaria* and *Pennisetum*. However, the evolutionary potential of even these appears to be limited, and Stebbins (1971: 179) makes the point that

> ... the hypothesis that apomictic complexes will not give rise to new genera or families, or even to characteristics not present in their sexual diploid ancestors, is very plausible. The principal support for this hypothesis lies in the fact that all known apomictic complexes are contained within the same genetic comparium, in that their range of variation is bounded by a group of sexual or facultatively apomictic species which are capable of hybridizing and exchanging genes with each other.

The debate on the evolutionary potential of asexual organisms has largely concentrated on these long-term consequences of asexuality. There is little that can be added to this debate. Consequently, this section will concentrate on a related issue, which concerns the capacity of asexual lineages to undergo genetic change in the short term. Basically, the issue for discussion is whether the short-term capacity of an asexual lineage for genetic change is different from that expected under the assumption that a mutation for asexuality has no effect other than to switch the pattern of reproduction from sexual to asexual. There have been few investigations of this issue and this section will consequently be brief. But there are some indications that the acquisition of at least some forms of asexuality may be accompanied by changes in both the frequency and the pattern of mutation. Three phenomena are of relevance here. The first is the positive association found between some forms of asexual reproduction, most notably parthenogenetic apomixis, and polyploidy. The second is the association between asexuality and hybridization. The third is the potential for genomic destabilization possessed by the family of meiotic mutations that have been identified as strong candidates for apomixis genes (Chapter 5).

Polyploidy can enhance the mutation rate per individual per generation. Thus if a constant number of mutations per haploid genome per generation is assumed, then it follows that a triploid will experience three mutations for every two experienced by a diploid. Manning and Dickson (1986) have argued that this effect of polyploidy is the main reason for its prevalence among asexual organisms, as it increases the mutation rate towards an optimum. I believe that this view is mistaken. It seems to predict that polyploidy should be both frequent and fairly evenly distributed among lineages that are obligately asexual (at least via the female function), and that its incidence should not be particularly affected by the pattern of asexual reproduction. But this is not the case. Polyploidy is not prevalent, for example, among asexual lower plants and animals or among asexual fungi (Burnett, 1975; Lewis, 1980; Lobban and Wynne, 1981; Mogie, 1986b), nor is it as common among automictic animals or among flowering plants reproducing by adventitious embryony as it is among parthenogenetically apomictic animals and plants. Rather, it is particularly associated with parthenogenetic apomixis and, as described in Chapter 5, this association is maintained irrespective of whether affected organisms are obligately or facultatively apomictic. This is not the distribution expected if the primary selective value of polyploidy is to help overcome the costs of uniparental inheritance – whether by optimizing the mutation rate per individual or by some other means. However, it is not an unexpected distribution if, as argued in Chapter 5, its primary role is to generate the conditions that favour the expression of the meiotic mutations that are responsible for the apomictic phenotype. This is not to say that polyploidy cannot have a beneficial effect on the mutation rate under some sets of circumstances. But if this is the case it appears that asexual organisms do not have a particularly high probability of encountering them. Otherwise, polyploidy would be spread more evenly across taxa exhibiting different types of asexual reproduction.

Irrespective of its primary selective value, a seemingly inescapable consequence of polyploidy is an enhanced mutation rate per individual per unit time. Given that many asexual organisms that do not require polyploidy for the expression of the asexual phenotype have not acquired it, it is reasonable to question whether organisms that do require it for this purpose find its enhancing effects on the mutation rate beneficial or detrimental. One obvious problem facing an asexual lineage which, through polyploidization, acquires an enhanced mutation rate is that there is an increase in the number of deleterious mutations per unit time as well as an increase in the number of advantageous mutations. Any gains in fitness resulting from the latter may be outweighed by costs resulting from the former. Polyploidy can help to buffer a lineage against an accumulation of deleterious mutations. But it is unclear whether this extra buffering capacity is generally sufficient to absorb all the extra deleterious mutations generated. To some extent, the net fitness effect of polyploidy will depend on the nature of the interactions between the different wild-type and mutant alleles at the affected loci. This problem has been investigated in computer simulations which looked at the consequences of an accumulation of deleterious mutations on a quantitative character in diploid and triploid asexual lineages (Mogie and Ford, 1988). The conditions used in these simulations were arbitrary and necessarily highly specified, but the results are worth reporting as they at least give a feel for the problem. Fitness was assumed to be positively correlated with the phenotypic value of the character. The details of the simulations are provided in Table 6.2. Table 6.2 shows that, under the specified set of conditions, polyploidy frequently, but by no means always, buffered an organism against the effects of accumulating deleterious mutations. Thus only under some circumstances was there a net benefit to its acquisition. The simulations showed that the buffering capacity of polyploidy can be dramatically affected by the pattern of interaction between alleles. The slowest decline in phenotypic value (and thus in fitness) caused by mutation was found in triploids when the loci controlling the character were initially heterozygous and the most advantageous of the two wild-type alleles included in the simulations was completely dominant to the less advantageous one. However, because triploids experience three mutations for every two experienced by diploids, polyploidy only retarded the decline in fitness of a lineage if the ratio of the number of mutations required for a given decline (diploids:triploids) was less than 0.67. It can be seen from Table 6.2 that polyploidy had no significant effect on this rate when alleles at homozygous loci interacted additively. The rate of decline increased when alleles at heterozygous loci interacted additively. For all other combinations polyploidy retarded the rate of decline in fitness, with the most dramatic decline being shown for heterozygous loci containing two dominant alleles and a recessive allele. The absence of a general advantage to polyploidy illustrated by the simulations reinforces the doubt expressed above about whether there is necessarily an overall advantage to an asexual lineage acquiring a higher mutation rate through polyploidy. Clearly, a much more rigorous approach needs to be taken to this problem, but if it turns out that there is no overall advantage to polyploidy then apomicts could be at a disadvantage, because of their requirement to retain polyploidy to enable the expression of the apomictic phenotype. These may suffer a cost of polyploidy, expressed as a decrease in evolutionary potential.

The enhancing effect of polyploidy on the mutation rate per individual per generation is unlikely to affect the relative frequencies of the different types of mutations produced (deletions, substitutions, inversions, translocations etc.). However, it is reasonable to argue that, even in the absence of any effects due to polyploidy, these frequencies, and indeed the mutation rate itself, may be affected by two other phenomena – the frequent association between asexuality and hybridization, and pleiotropic effects of genes for asexuality.

Hybridization can enhance the mutation rate, including that of chromosome mutations. For example, Sturtevant (1939) found that the mutation rate in the progeny of a cross between two races of *Drosophila pseudoobscura* was far higher than expected, and Woodruff *et al.* (1979) have made similar observations for *D. melanogaster*. Similarly, Gustafsson (1947) found that the mutation rate in pure lines of *Hordeum* was far lower than that found in their sexual hybrids. With respect to chromosome mutations, Giles (1940) reported that the rate of spontaneous chromosome breakage during male meiosis in a natural interspecific hybrid of *Tradescantia* was, on average, three times that of several pure *Tradescantia* species, and that the hybrids exhibited bridges, fragments and deletions. Walters (1950) reported a high frequency of bridges and fragments at both anaphases of meiosis in sterile but often highly vigorous hybrids of *Bromus trinii* and *B. maritimus*. Unlike their hybrids, these species exhibited completely regular meioses. Bridges and fragments at anaphase have also been reported in interspecific hybrids of *Allium* (Emsweller and Jones, 1938) and

Table 6.2 The number of mutations required to reduce the phenotypic value of a quantitative character to 90% of its initial value, and the relative rate of approach to this value (number of mutations required in diploids/number required in triploids) in diploid and triploid apomictic lineages. Different lineages exhibit different starting allele frequencies and different modes of interaction between alleles within a locus. If it is assumed that triploids experience three mutations for every two mutations experienced by diploids, then relative rates of approach greater than 0.67 indicate that triploids experience a 10% reduction sooner than diploids. The opposite is the case if rates fall below 0.67.

The character is assumed to be influenced by 240 loci, each of which can be homozygous for either of two types of non-mutant allele (A or a) or heterozygous. Alleles within a locus are allowed either to interact additively with respect to fitness, with complete dominance of A-types over a-types, or with overdominance. The pattern of interaction is the same for all loci within a simulation but varies between simulations. Mutant alleles are introduced at random during each simulation. A site can experience more than one mutational event, but back-mutations are not allowed. The phenotype is determined by assigning values to each of the three allele types, allowing the calculation of locus values and, by summation over all loci, phenotypic values. In this simulation, A-type alleles are given mean phenotypic values which are double those of a-type alleles and ten times those of mutant alleles. Fitness is assumed to be positively correlated with phenotypic values. Based on Table 4 of Mogie and Ford (1988)

| Initial genotype of diploids (%) | | | Locus behaviour (%) | | | Number of mutations required to reach 90% of initial phenotypic value (+/−SD) ((xxx) = initial genotype of triploids) | | Relative rate of approach to 90% (2n/3n) |
AA	Aa	aa	Add.	D/R	H.A.	Diploid	Triploid	
0	100	0	0	0	100	42.6 +/− 0.97	106.3 +/− 11.87 (AAa)	0.40***
							107.4 +/− 18.20 (Aaa)	0.40***
0	100	0	0	100	0	74.2 +/− 10.67	312.7 +/− 24.78 (AAa)	0.24***
							118.9 +/− 21.49 (Aaa)	0.62***
100	0	0	100	0	0	57.3 +/− 2.06	86.5 +/− 2.17 (AAA)	0.66ns
0	100	0	100	0	0	70.6 +/− 5.87	89.1 +/− 5.36 (AAa)	0.79***
							90.8 +/− 5.63 (Aaa)	0.78***
0	0	100	100	0	0	68.3 +/− 3.16	98.8 +/− 2.53 (aaa)	0.69ns

*** P<0.001 (Chi-squared test).

ns = no significant difference between diploids and triploids in the rate of approach to 90% of initial phenotypic value.

Add. = alleles interact additively.

D/R = allele A completely dominant to allele a.

H.A. = overdominance

Crepis (Muntzing, 1934) and in intergeneric hybrids of *Elymus* and *Agropyron* (Sadasivaiah and Weijer, 1981).

It is unclear why hybridization can have this effect. With respect to intraspecific hybridization, Sturtevant (1939), and later Thompson and Woodruff (1978) and Woodruff and Thompson (1980), have argued that the mutation rate is genetically suppressed in many species, and that the genetic basis of suppression may vary between populations. Hybridization between populations could disrupt the coadapted gene complexes which had suppressed the effects of mutator genes. This argument can be easily extended to the interspecific case. Whatever the genetic basis for the hybridization-induced enhancement of the mutation rate, it is a clearly identified phenomenon which could have considerable consequences for the evolutionary potential of asexual taxa as many of these are hybrids. Indeed, the importance of hybridization in the initiation of asexual (agamosporous) homosporous lineages has been emphasized in Chapter 3, and many heterosporous generative and aposporous apomicts are clearly of hybrid origin. It is therefore of some interest that there are several reports of apomicts exhibiting a high rate of chromosome mutation. Thus Gentcheff and Gustafsson (1940) observed up to nine or ten bridges per anaphase plate during meiosis in tetraploid, generatively apomictic *Hieracium robustum* and *H. amplexicaule*. Bridges have also been observed during male meiosis in the generative apomicts *Calamagrostis purpurea* and *C. lapponica* (Nygren, 1946), *Dichanthium annulatum* (Reddy and D'Cruz, 1969) and *Poa alpina* (Muntzing, 1940), in the aposporous apomicts *Ranunculus auricomus* (Rousi, 1955; Marklund and Rousi, 1961) and *Hierochloe alpina* (Weimarck, 1973), and in apomicts of the grass tribe Paniceae (Chatterji and Timothy, 1969; Nath *et al.*, 1970). Mitotic chromosome bridges have been observed in *Taraxacum* (Mogie, 1982). Here, apomicts exhibited a higher frequency of mitotic anaphase bridges than sexual taxa although, interestingly, a raw, artificial sexual hybrid also showed a high incidence of bridge formation. These *Taraxacum* data are summarized in Table 6.3.

Further evidence that *Taraxacum* apomicts can experience chromosome instability comes from observations of a morphologically distinct satellited chromosome, which is found in 20 of the 33 taxonomic sections for which reliable cytological data are available (reviewed by Mogie and Richards, 1983). In nine sections this chromosome type is known to be structurally stable and it occurs at the rate of one per haploid genome. In six others it is known to be structurally unstable, there being interspecific, intraspecific and even intra-individual variation in its frequency. This variation is typically due to the disappearance of a part or the whole of the satellite segment, rather than of the entire chromosome, the genome remaining euploid. It appears that the

Table 6.3 The percentage of cells showing bridges at anaphase of mitosis in sexual and asexual *Taraxacum* species and hybrids

Taxon	Breeding scheme	Percentage of root-tip cells with anaphase bridges
T. pseudohamatum	obligately apomictic	10.5
T. unguilobum	obligately apomictic	6.6
T. lacistophyllum	obligately apomictic	3.8
T. alacre	obligately sexual, outbreeder	0.0
T. bessarabicum	obligately sexual, inbreeder	1.0
T. brevifloroides × T. oliganthum	sexual hybrid	8.1

satellite is either being deleted from the genome or is being hidden within the body of a chromosome following structural rearrangement. Richards (1989) has carried out a detailed investigation of the capacity of this chromosome to undergo structural rearrangement in four individuals of one sexual taxon and in five individuals of two apomictic taxa. No rearrangements were observed in any of the sexual individuals or in two of the apomicts analysed. But between 10 and 25% of the satellited chromosomes in the remaining three apomicts had undergone structural rearrangement (Table 6.4). These three individuals were drawn from both apomictic taxa, and each of these taxa included an individual which appeared to exhibit stable satellited chromosomes (Table 6.4). The relationship between the two *T. brachyglossum* individuals is unknown, but the three individuals of *T. pseudohamatum* were produced apomictically from the same mother and would thus presumably have inherited the same tendency for chromosome structural instability. That they did not appear to exhibit this tendency equally makes interpretation of the results difficult but, overall, the results do appear to confirm that this chromosome can be structurally unstable, especially when it is part of an apomictic chromosome complement. There is no reason to assume that the frequency of structural rearrangement of this chromosome is any different from that of the remainder of the chromosome complement, indicating a very high rate of chromosome mutation in some apomictic members of this genus.

As in the case of polyploidy enhancing the mutation rate, and for the same reason, it is not clear whether a hybridization-induced enhancement will be of net benefit or cost to an asexual lineage. However, if detrimental, its consequences may not be as severe, on

Table 6.4 A summary of Richards' (1989) data on the frequency of structural rearrangement of the satellited chromosome of *Taraxacum*. The four sexual individuals are half-sibs. The three *T. pseudohamatum* apomicts are full sibs. The relationship between the two apomicts of *T. brachyglossum* is not known

Taxon	Breeding scheme	Individual Code	No. satellited chromosomes examined	No. showing structural rearrangement
T. sect. *Vulgaria*	sexual	F	45	0
		H	21	0
		J	32	0
		K	15	0
T. pseudohamatum	apomict	A	50	5
		C	20	0
		D	20	5
T. brachyglossum	apomict	E	16	3
		G	21	0

average, as those of polyploidy. The enhancing effect of polyploidy on the mutation rate of individuals is predictable and must be largely unavoidable, as this rate is a function of the amount of DNA. The probability of a lineage that is being deleteriously affected by this rate accumulating mutations that will lower the rate per haploid genome so that the rate per individual approaches an optimum must be low. Consequently, if polyploidy is central to the expression of the asexual phenotype – as it appears to be in almost all parthenogenetic apomicts – affected lineages must bear any associated costs. In contrast, the enhancing effect of hybridization on the mutation rate may, on occasion, be as pronounced or more pronounced as that of polyploidy, but it is not nearly so inevitable. In many cases it is absent, and when it is present its level can vary considerably among affected individuals (e.g. Giles, 1940), providing conditions under which natural selection will favour those lineages with individual mutation rates closest to the optimum.

The third phenomenon that may contribute to an enhanced mutation rate in some asexual organisms are the meiotic mutations implicated in the control of parthenogenetic apomixis. It was mentioned in Chapter 5 that a number of these mutations affect mitosis as well as meiosis, and that these mitotic effects can include an enhanced rate of mitotic recombination and an enhanced mutation rate, especially of chromosome mutations (Baker *et al.*, 1976; Lewin, 1980). Possibly the enhanced mutation rates recorded in some apomicts, which are described above, may result from pleiotropic effects of these genes rather than from hybridization. There is no way to distinguish clearly between these possibilities at present, but the potential of these genes as enhancers of the mutation rate should not be over-looked.

It is clear that there are various ways in which the mutation rates of asexual organisms can be enhanced. However, it is far from clear whether asexual organisms will obtain costs or benefits from such enhancements. Nevertheless, there are some indications from electrophoretic studies involving generatively apomictic *Taraxacum* species that an enhanced mutation rate may impart considerable evolutionary potential. Much of this information has been obtained from investigations of members of the Section *Hamata*, a section which is known to contain taxa that exhibit high rates of chromosome mutation (Mogie, 1982; Mogie and Richards, 1983; Richards, 1989). Reviewing this information offers a suitably tentative means of concluding this chapter.

Ford and Richards (1985) and Mogie (1985) have detected electrophoretic (enzymic) variants among samples of only 12–25 sibs obtained asexually from obligately apomictic *Taraxacum* species. Similarly, King and Schaal (1990) have detected heritable variation within a lineage in mid-repetitive rRNA-encoding DNA (rDNA) and in single copy *Adh1* (the genes encoding alcohol dehydrogenase 1). Ford and Richards (1985), Mogie (1985) and Van Oostrum *et al.* (1985) have also demonstrated that electrophoretic variation is common within *Taraxacum* populations. For example, Ford and Richards (1985) collected individuals of 10 apomictic species from an area of stable dune grassland measuring only 100 m². Seven of these species were represented by two or more individuals, and five of these exhibited more than one esterase zymogram (Table 6.5). These analyses of sibs and natural populations indicate that some, at least, of the variants within a species share a common ancestry.

Table 6.5 The number of individuals assayed for esterase, and the number of zymograms recorded, for seven apomictic *Taraxacum* species collected from a 100 m² area of stable dune grassland. Data from Ford and Richards (1985); based on Table 2 of Mogie and Ford (1988)

Species	No. of individuals assayed	No. of variants recorded
T. proximum	2	1
T. unguilobum	8	2
T. fulviforme	12	2
T. subnaevosum	10	1
T. nordstedtii	9	2
T. brachyglossum	12	3
T. lacistophyllum	41	4

Given the apparently high mutation rates and high incidence of electrophoretic variation within *Taraxacum*, it is pertinent to consider whether these phenomena are sufficient to enhance the evolutionary potential of apomicts of this genus. An indication that this may be the case is provided by a study undertaken into the frequency and distribution of electrophoretic variants within Section *Hamata* (Mogie, 1985). This section comprises 24 species, all of which are triploid and appear to be obligately apomictic. The section occurs over much of the western seaboard of Europe, in 12 countries. One species, *T. hamatum* is recorded in all 12 countries, and another, *T. hamatiforme*, is absent from only one. Øllgaard (1983) considers these two species to be the oldest representatives of the section. The 24 species are morphologically and karyologically similar to one another, but differ considerably in each of these characteristics from other members of the genus. Most notable is their possession of only two satellited chromosomes – taxa in the more closely related sections tend to have one per haploid genome (Mogie and Richards, 1983). Thus hybridization between members of this section and taxa from other sections is very unlikely to produce variants with karyotypes or morphologies typical of Section *Hamata*. These observations led Mogie and Richards (1983) to suggest that the members of this section could be descended from a common apomictic ancestor. The electrophoretic study offers some validity to this conclusion. Briefly, 60 individuals representing eight of the 24 species were assayed for esterase. These individuals were obtained from sites throughout Britain and from Denmark and

Table 6.6 The distribution of esterase zymograms among eight obligately apomictic species of the section *Hamata* of the genus *Taraxacum*. Data from Mogie (1985)

Species	C	IND	ZYM	a	b	c	d	e	f	g	h
T. hamatum	12	8	4	5	1	–	1	1	–	–	–
T. hamatiforme	11	8	5	4	1	1	–	–	1	–	1
T. hamiferum	8	2	2	1	–	–	–	–	–	–	1
T. quadrans	7	14	3	12	–	–	–	–	1	1	–
T. atactum	7	5	1	5	–	–	–	–	–	–	–
T. kernianum	5	2	1	2	–	–	–	–	–	–	–
T. pseudohamatum	4	18	4	7	1	5	–	–	–	5	–
T. boekmanii	4	3	1	3	–	–	–	–	–	–	–
No. of individuals				39	3	6	1	1	2	6	2
No. of species				8	3	2	1	1	2	2	2

C, number of countries in which the species has been recorded.
IND, number of individuals assayed.
ZYM, number of zymograms recorded.
a-h, zymogram types.
From Table 3 of Mogie and Ford (1988).

Sweden. The data are summarized in Table 6.6. In all, eight different zymograms were observed, the commonest of which, zymogram a, was found in 39 (65%) of the accessions, in all eight species, and throughout the sample area. Six of the species exhibited more than one zymogram, and the two most widespread and putatively ancient taxa (*T. hamatum* and *T. hamatiforme*) were also the most variable with respect to this character. Both of these observations support the conclusion drawn from morphological and karyological studies that these species are very closely related, and they lend weight to the suggestion that they may all be derived from a single apomictic ancestor. If this is the case then it is clear that *Taraxacum* apomicts can show considerable evolutionary potential.

This possibility was further investigated in a study that examined morphological variation in obligately apomictic *T. pseudohamatum* (Mogie, 1985). It can be seen from Tables 6.3, 6.4 and 6.6 that this species is variable with respect to esterase zymograms and exhibits structurally unstable chromosome complements. Six morphometric characters were measured in seven accessions grown under uniform conditions. The accessions had been collected from sites throughout England and Wales. Three of the accessions were similar for all characters, but considerable differences in one or more characters were observed in all other pairwise comparisons. One observation of considerable interest was that accessions collected from sites that were widely separated tended to be more different than accessions collected from neighbouring sites. A regression analysis showed this trend to be highly significant ($P < 0.001$). These data are summarized in Table

Table 6.7 (a) The number of characters by which seven *T. pseudohamatum* accessions exhibited significant differences. Six morphometric characters were used in the analysis. These were capitulum diameter, the widths of the lowermost and uppermost exterior involucral bracts, achene number, lobe number of the longest leaf and the ratio between the length of the longest leaf and the width, at its widest point, of its terminal lobe. (b) The distance (km) separating the sites from which the seven accessions were collected. From Table 2 of Mogie (1985)

	Accession		Accession						Accession		Accession				
	Accession	84	88	121	65	98	20		Accession	84	88	121	65	98	20
(a)	84							(b)	84						
	88	0							88	0					
	121	0	0						121	8	8				
	65	2	1	1					65	250	250	250			
	98	3	2	1	3				98	150	150	150	200		
	20	1	2	4	3	4			20	400	400	400	240	290	
	53	3	3	4	4	5	3		53	600	600	600	400	460	200

6.7. It is likely that the distance separating two accessions is positively correlated with the length of time that they have been separated. The observation that widely separated accessions are more diverse than neighbouring ones may not therefore be viewed as surprising, except that *T. pseudohamatum* is obligately apomictic. Thus overall, there is a considerable body of evidence that *Taraxacum* apomicts can have enhanced mutation rates and that these may not necessarily be detrimental but can lead to considerable morphological and biochemical divergence. There is no obvious reason for believing that *Taraxacum* apomicts are unique in this respect and thus for believing other than that apomicts, and possibly hybrid or polyploid organisms with other methods of asexual reproduction, may have unexpectedly high evolutionary potentials.

Chapter 7

Reflections

Every so often, a problem arises that arouses the interest of biologists from a number of disciplines. The problem of the evolution, maintenance and control of asexual reproduction is a leading example, having stalked the highways and byways of taxonomy, cytology, genetics, embryology, ecology and evolutionary biology for many decades. The expertise of these disciplines must continue to probe the biology of asexuality if we are to achieve a thorough understanding of it. However, it is debatable whether this level of understanding will be approached without a change in the emphasis and direction of some aspects of this research.

To some extent, some of the major thrusts of this research have been too narrowly directed. Evolutionary biologists have tended to view asexuality as though it is an obligate condition and, perhaps too exclusively, have directed their gaze between generations for an understanding of its costs, benefits and distribution; ecologists have been too ready to interpret the secondary characteristics with which asexuality is sometimes associated (e.g. polyploidy) as primarily ecological adaptations; and geneticists have been over-confident in assuming that an understanding of the types of changes that must be undergone during a transition from sexual to asexual reproduction can be gleaned simply from an examination of those lineages that have successfully completed this task. Conversely, the importance of developmental and phylogenetic constraints, and of the male function in cosexual forms, has not been sufficiently appreciated, and the search for parsimony in genetic models has not always been conducted with vigour.

The main aspects of asexual reproduction that have been investigated in this book are its taxonomic distribution, its establishment and maintenance, and the nature of its control. These are inextricably intertwined. One message above all has emerged from these investigations. This is that some taxa are pre-adapted to evolve asexual reproduction, and that it is primarily, perhaps exclusively, from this group that asexual lineages have arisen. That is, taxa that are not pre-adapted to evolve asexuality do not appear simply to take longer than pre-adapted taxa to evolve it; they appear to fail to evolve it. Thus sex may be pre-eminent because there is a cohort of phylogenetic, ontogenetic and reproductive constraints that has prevented most taxa from evolving asexual reproduction.

These constraints confront the individual or lineage as it undertakes the evolutionary journey from sexuality to asexuality. They reflect a problem of magnificent proportions: this is that the transition from sexual to asexual reproduction requires numerous changes to an individual's reproductive biology that are individually highly deleterious. In the adaptive topography, sexual reproduction may occupy one peak and asexual reproduction another, possibly higher, peak but these are separated by valleys of sterility or of deleterious genome loss or gain. Thus the transition from sexual to asexual reproduction in homosporous organisms, where cost-benefit analysis indicates that asexual reproduction provides a selectively advantageous alternative to female sterility but not to sexual reproduction (Chapter 3), requires the acquisition of female sterility, the avoidance of meiotic reduction during meiospore production, the acquisition of apogamy, and the ecological isolation of newly arisen asexual mutants from their more fecund sexual conspecifics. The transition in heterosporous organisms, where cost-benefit analysis indicates that asexual reproduction provides a selectively advantageous alternative to sexual reproduction requires, at a minimum, the avoidance of meiotic reduction during megaspore production and of fertilization following egg maturation, and the acquisition of the ability to initiate embryos parthenogenetically; in cosexual forms it may also require the retention of a sexual male function and, in flowering plants, it may require the acquisition of the ability to initiate embryo sacs aposporously and the retention of polar nucleus fertilization to ensure endosperm development. Clearly, acquiring only the capacity to avoid meiotic reduction would be deleterious, as would acquiring only the capacity to initiate embryos parthenogenetically or only the capacity to avoid fertilization. Rather than resulting in the generation of a successful pattern of asexual reproduction these would, respectively, result in the fertilization of unreduced eggs, haploidy and female sterility.

It seems that the only way organisms have been able to overcome the problem of the components of asexuality being individually deleterious is to bring them together more-or-less simultaneously and in phenotypically expressible form. But it seems from the restricted and uneven taxonomic distribution of asexuality and from the relative rarity of asexual reproduction compared with sexual reproduction that most organisms have been prevented by their ontogenies or by aspects of their (sexual) reproductive ecologies from adopting this strategy. This is why I argue that pre-adaptation may well be an absolute requirement for the evolution of asexuality. Stated very briefly, the major problem organisms face in evolving asexuality is that it is extremely unlikely that the different components of asexuality will be brought together more-or-less simultaneously and in phenotypically expressible form if their accumulation requires the acquisition of several independent, and necessarily phenotypically dominant, mutations. And yet organisms that are not pre-adapted for the evolution of asexuality are faced with this task. The period available for the transition is just too short to expect such organisms to complete it. (For example, a lineage that acquires a gene for the avoidance of meiotic reduction but not one for the avoidance of fertilization will experience highly deleterious ploidy levels after only a few generations.) Pre-adapted organisms exhibit ontogenies or patterns of sexual reproduction which enable them to complete the transition from sexual to asexual reproduction without having to acquire numerous independent mutations.

The pre-adaptions take different forms in different taxa. Chapter 3 argues that, among homosporous taxa, the major pre-adaptation appears to be a tendency to undergo hybridization, as this can simultaneously induce two of the components required by homosporous organisms if they are to complete the transition from sexual to asexual reproduction. These are female sterility and ecological isolation. The importance of this pre-adaptation is very clear, as asexuality is found only in that group of homosporous organisms – the ferns – where hybridization is common; it is absent from the bryophytes, where hybridization is rare and where, consequently, female sterility and ecological isolation would have to be achieved by mutation. Among heterosporous taxa, pre-adaptation appears to be associated with a number of characters and characteristics. One concerns the timing of pollination relative to egg maturation. It was shown in Chapter 3 how differences in this characteristic between gymnosperms and flowering plants potentially provide the latter with a strategy to avoid the fertilization of the egg that is not available to the former. Flowering plants have the option of developing eggs precociously (before pollen is released) in order to avoid fertilization, as pollination occurs at about the time of egg maturation. They do not have to prevent pollination in order to avoid fertilization. In contrast, precocity is not an option available to gymnosperms, as pollination occurs long before egg maturation in this group. These must prevent pollination if they are to avoid fertilization. This difference between the two groups appears to be very important, as no asexual flowering plant avoids fertilization by preventing pollination. They effectively simply delay pollination

by maturing eggs precociously. This suggests that the latter strategy
is much easier to acquire than the former which will almost certainly
require the accumulation of several mutations. Evidence presented in
Chapter 3 that this is the case, argues that precocity can simply be a
pleiotropic consequence of the avoidance of meiotic reduction. Thus
the timing of pollination relative to egg maturation is pre-adaptive in
flowering plants. It removes a barrier to the evolution of asexuality –
namely the requirement to acquire a mutation if fertilization is to
be avoided – that confronts gymnosperms and that has effectively
precluded its evolution in this group. However, it is argued in Chapter
5 that precocious egg maturation is only one of several pre-adaptations
that are required by flowering plants if a successful transition from
sexual to asexual reproduction is to be achieved. Indeed, even precocity may not be pre-adaptive in many taxa as the degree of precocity
induced by the avoidance of meiotic reduction will vary between taxa
and in only a proportion of these will it be at or about the level that
is conducive to a viable pattern of asexual reproduction. But even
when it reaches this level, asexuality is very unlikely to arise unless
other components of the transition can be acquired without mutation.

There is overwhelming evidence that the events associated with
megaspore production in apomicts (the avoidance of meiotic reduction
in generative apomicts, or the degeneration of the megaspore or its
products in aposporous apomicts) will require the acquisition of a
mutation (Chapter 5). Thus it is unreasonable to consider that pre-
adaptation exists with respect to this character. But there is strong
evidence that parthenogenesis in flowering plants need not require a
mutation, being an innate capacity of many organisms. Thus it can
be reasonably argued that flowering plants that show an innate
capacity for parthenogenesis and that, on receipt of a meiotic
mutation, will generate eggs sufficiently precociously to enable this
capacity to be realized, are pre-adapted to evolve asexual reproduction.
Asexual reproduction will be achieved by acquiring only a single meiotic mutation as this induces precocious egg maturation which allows
the innate parthenogenetic capacity of the egg to be expressed.

The role of pre-adaptation becomes more and more evident the more
deeply the control of asexuality is probed. Thus the innate capacity
for parthenogenesis exhibited by many flowering plants appears to
be a pre-adaptation born out of a fortuitous phylogenetic shift that
accompanied the evolution of heterospory from homospory. Thus,
from Bell's (1979a,b) studies of egg division in homosporous ferns, it
appears that the ancestral egg type among land plants lacks parthenogenetic potential (Chapter 3). However, it seems that the eggs of heterosporous taxa are homologous to mitotically competent precursors of
the ancestral egg type. This mitotic competence appears to have been

retained by many taxa, being translated into parthenogenetic potential, as these precursors adopted the role of the egg.

The problems concerning the evolution of asexual reproduction comprise only a fraction of the problems associated with asexuality. Others concern its establishment, maintenance and spread. It has been argued, in Chapters 3 and 4, that these aspects of asexuality have been profoundly influenced in cosexual forms by the male function. I argue that this function, by enabling individuals that are asexual via their female function to reproduce sexually as male parents, has had a profound effect on diverse aspects of asexuality in cosexual forms. Its major importance is that it provides the mechanism by which genes for asexuality can be inserted into new genomes, some of which will be far superior to the genomes the genes originally arose in. The importance of this phenomenon cannot be overemphasized. Indeed, it may be that because of it the genomes of cosexual asexual taxa will, on the whole, be superior to those of dioecious asexual taxa; an asexual mutation arising in a female of a dioecious taxon is restricted to the one genome; it is unlikely that this genome will be the best that the population gene pool can offer. Similarly, because of the male function, any genetically determined ecological edge that sexual individuals may happen to have over asexual conspecifics may be lost when they confront each other. The genes conferring ecological superiority will quickly become incorporated into the asexual component of the taxon as asexual individuals fertilize the eggs of sexual competitors, producing new asexual lineages in the process. This argument is expanded in Chapter 4, to illustrate how the male function may provide homosporous taxa with the means comprehensively to overcome the cost of asexual reproduction compared with sexual reproduction, and how it may have played the major role in determining the geographic distribution of related sexual and asexual forms.

Given that this book is about asexual reproduction, which is fundamentally a characteristic of the female function, my highlighting of the role of the male function may appear inappropriate. But its role in the establishment, maintenance and geographical distribution of asexual reproduction in cosexual taxa does appear to have been underappreciated. So too has the role of pre-adaptation in the evolution of asexual reproduction. As a way of concluding this book I will point out again that by underestimating the importance of these two phenomena the importance of cost-benefit analysis when applied to the issue of the maintenance of sexual reproduction may have been overestimated. The pre-eminence of sexual reproduction may indicate no more than that most taxa are not pre-adapted, and thus are unable, to evolve asexual reproduction. This has important implications for the ongoing debate, discussed in Chapter 6, about the selective value of genetic recombination, and therefore of sex: if most sexual taxa have never been exposed to a challenge from asexuality because they are not pre-adapted to evolve it, the utility of invoking adaptive arguments to explain the pre-eminance of sexual reproduction is called into question. That is, asexuality may be rare compared with sexuality because it arises rarely, not because it is in some way usually inferior to sexual reproduction. We may need to go no further than this to understand why most organisms are sexual. Clearly, many more investigations into the evolution of asexual reproduction and the ecology of asexual organisms need to be undertaken before we will be in a position to test this hypothesis. But it is one that needs to be kept clearly in view.

References

Abdullah, N. F. and Charlesworth, B. (1974) Selection for reduced crossing over in *Drosophila melanogaster*. *Genetics*, **76**, 447–51.

Åkerberg, E. and Bingefors, S. (1953) Progeny studies in the hybrid *Poa pratensis* × *Poa alpina*. *Hereditas*, **39**, 125–36.

Allard, R. W. (1963) Evidence for genetic restriction of recombination in the lima bean. *Genetics*, **48**, 1389–95.

Allen, G. S. (1942) Parthenocarpy, parthenogenesis and self-sterility in Douglas-fir. *J. Forestry*, **40**, 642–4.

Andersson, E. (1947) A case of asyndesis in *Picea abies*. *Hereditas*, **33**, 301–47.

Antonovics, J. and Ellstrand, N. C. (1984) Experimental studies of the evolutionary significance of sex. I. A test of the frequency-dependent selection hypothesis. *Evolution*, **38**, 103–15.

Apirion, D. (1966) Recessive mutants at unlinked loci which complement in diploids but not in heterokaryons of *Aspergillus nidulans*. *Genetics*, **53**, 935.

Ashton, N. W. and Cove, D. J. (1977) The isolation and preliminary characterization of auxotrophic and analogue resistant mutants of the moss, *Physcomitrella patens*. *Mol. Gen. Genet.*, **154**, 87–95.

Asker, S. (1980) Gametophytic apomixis: elements and genetic regulation. *Hereditas*, **93**, 277–93.

Atkinson, L. R. (1975) The gametophyte of five old world Thelypteroid ferns. *Phytomorphology*, **25**, 38–54.

Babcock, E. B. and Stebbins, G. L. (1938) The American species of *Crepis*. Their interrelationships and distribution as affected by polyploidy and apomixis. Carnegie Inst. Washington, Publ. No. 504.

Baker, B. S., Carpenter, A. T. C., Esposito, M. S., Esposito, R. E. and Sandler, L. (1976) The genetic control of meiosis. *Ann. Rev. Genet.*, **10**, 53–134.

Baker, B. S. and Hall, J. C. (1976) Meiotic mutants: genetic control of meiotic recombination and chromosome segregation. in *Genetics and Biology of Drosophila 1a*, (eds E. Novitski and M. Ashburner) Academic Press, New York, pp. 351–434.

Baker, H. G. (1955) Self-compatibility and establishment after 'long-distance' dispersal. *Evolution*, **9**, 347–9.

Baker, H. G. (1965) Characteristics and modes of origin of weeds, in *The Genetics of Colonizing Species* (eds H. G. Baker and G. L. Stebbins) Academic Press, New York, pp. 147–72.

Baker, H. G. (1966) The evolution, functioning and breakdown of heteromorphic incompatibility systems. *Evolution*, **20**, 349–68.

Baker, H. G. (1967) Support for Baker's Law as a rule. *Evolution*, **21**, 853–6.

Barker, W. W. (1966) Apomixis in the genus *Arnica* (Compositae). PhD thesis, University of Washington, Seattle.

Barlow, P. W. (1989) Meristems, metamers and modules and the development of shoot and root systems. *Bot. J. Linn. Soc.*, **100**, 255–79.

Bashaw, E. C. (1962) Apomixis and sexuality in bufflegrass. *Crop Sci.*, **2**, 412–15.

Battaglia, E. (1946) Ricerche cariologiche ed embriologiche sul genre *Rudbeckia* (Asteraceae). VI. Apomissia, in *Rudbeckia speciosa*. *Nuovo G. Bot. Ital. (NS)*, **53**, 27–69.

Battaglia, E. (1950) L'alterazione della meiosi nella reproduzione apomittica di *Erigeron karwinskianus* DC. var *mucronatus* DC. *Caryologia*, **2**, 165–204.

Battaglia, E. (1987) Embryological questions: 10. Have the expressions 'Polar Nuclei' and 'Secondary Nucleus' been rightly established? Appendix: Hofmeister W. (1847), 'Untersuchungen . . . Bei . . . Oenothereen'. *Atti Della Societa Toscana Di Scienze Naturali*, **94**, 127–50.

Battaglia, E. (1988) Embryological questions: 13. Can the atypical 4-nucleate embryo sacs of Tamarix assigned to the Plumbagella type (haploid egg) be reinterpreted as new Tamarix types (diploid egg)? *Atti Soc. Tosc. Sci. Nat.*, **95**, 69–82.

Battaglia, E. (1989) The evolution of the female gametophyte of angiosperms: an interpretative key. (Embryological questions: 14). *Annali di Botanica.*, **47**, 7–144.

Bayer, R. J. and Stebbins, G. L. (1983) Distribution of sexual and apomictic populations of *Antennaria parlinii*, *Evolution*, **37**, 555–61.

Beaman, H. G. (1957) The systematics and evolution of *Townsendia* (Compositae). Contrib. Gray Herbarium (Harvard) No. 183.

Beatty, R. A. (1967) Parthenogenesis in vertebrates. in *Fertilization: Comparative Morphology, Biochemistry and Immunology, vol. 1*. (ed. C. B. Mertz and A. Monroy) Academic Press, New York, pp. 413–40.

Beatty, R. A. (1975) Genetics of animal spermatozoa. Gamete Competition, in *Plants and Animals* (ed. D. L. Mulcahy) North Holland, Amsterdam, pp. 61–8.

Beck, C. B. and Wight, D. C. (1988) Progymnosperms, in *Origin and Evolution of Gymnosperms* (ed. C. B. Beck) Columbia University Press, New York, pp. 1–84.

Bell, G. (1982) *The Masterpiece of Nature. The Evolution and Genetics of Sexuality*. Croom Helm, London.

Bell, P. R. (1979a) The contribution of ferns to an understanding of the life cycles of vascular plants, in *The Experimental Biology of Ferns* (ed. A. F. Dyer) Academic Press, London, pp. 57–85.

Bell, P. R. (1979b) Gametogenesis and fertilization in ferns. in *The Experimental Biology of Ferns* (ed. A. F. Dyer) Academic Press, London, pp. 471–503.

Bengtsson, B. (1985) Biassed gene conversion as the primary function of recombination. *Genet. Res.*, **47**, 77–80.

Bennett, M. D. (1972) Nuclear DNA content and minimum generation time in herbaceous plants. *Proc. R. Soc. Lond. [Biol]*, **181**, 109–35.

Bennett, M. D. and Smith, J. B. (1972) The effects of polyploidy on meiotic duration and pollen development in cereal anthers. *Proc. R. Soc. Lond. [Biol]*, **181**, 81–107.

Bergman, B. (1951) On the formation of reduced and unreduced gametophytes in the females of *Antennaria carpatica*. *Hereditas*, **37**, 501–18.

Bergman, B. (1952) Asyndesis in macrosporogenesis of diploid, triploid and tetraploid *Chrysanthemum carinatum*. *Hereditas*, **38**, 83–90.

Bernstein, H. (1977) Germ line recombination may be primarily a manifestation of DNA repair processes. *J. Theor. Biol.*, **69**, 371–80.

Bernstein, H. (1983) Recombinational repair may be an important function of sexual reproduction. *BioScience*, **33**, 326–31.

Bernstein, H., Byers, G. S. and Michod, R. E. (1981) Evolution of sexual reproduction: importance of DNA repair, complementation and variation. *Am. Nat.*, **117**, 537–49.

Bernstein, H., Byerly, H. C., Hopf, F. and Michod, R. E. (1984) Origin of sex. *J. Theor. Biol.*, **110**, 323–51.

Bernstein, H., Byerly, H. C., Hopf, F. and Michod, R. E. (1985a) The evolutionary role of recombinational repair and sex. *Int. Rev. Cytol.*, **96**, 1–28.

Bernstein, H., Byerly, H. C., Hopf, F. and Michod, R. E. (1985b) Genetic damage, mutation and the evolution of sex. *Science*, **229**, 1277–81.

Bernstein, H., Hopf, F. A. and Michod, R. E. (1988) Is meiotic recombination an adaptation for repairing DNA, producing genetic variation, or both?, in *The Evolution of Sex* (eds R. E. Michod and B. R. Levin) Sinauer Associates, Sunderland, MA, pp. 139–60.

Bhandari, N. N. (1984) The microsporangium, in *Embryology of Angiosperms* (ed. B. M. Johri) Springer-Verlag, Berlin, pp. 53–121.

Bierzychudek, P. (1981) Pollinator limitation of plant reproductive effort. *Am. Nat.*, **117**, 838–40.

Bierzychudek, P. (1987a) Resolving the paradox of sexual reproduction: a review of experimental tests, in *The Evolution of Sex and its Consequences* (ed. S. C. Stearns) Birkhauser Verlag, Basel, pp. 163–74.

Bierzychudek, P. (1987b) Patterns in plant parthenogenesis, in *The Evolution of Sex and its Consequences* (ed. S. C. Stearns) Birkhauser Verlag, Basel, pp. 197–217.

Bocher, T. W. (1951) Cytological and embryological studies in the amphiapomictic *Arabis Holbeollii* complex. *K Dan Vidensk Selsk Biol. Skr. VI*, **7**, 1–59.

Bogdan, A. V. (1963) A note on breeding behaviour of *Panicum maximum*. *E. Afr. Agr. J.*, 206–17.

Bold, H. C. and Wynne, M. J. (1978) *Introduction to the Algae*. Prentice-Hall, New Jersey.

Bopp, M. (1968) Control of differentiation in fern allies and bryophytes. *Annu. Rev. Plant Physiol.*, **19**, 361–77.

Bostock, C. J. and Sumner, A. T. (1978) *The Eukaryotic Chromosome*. North Holland, Amsterdam.

Bouman, F. (1984) The ovule. in *Embryology of Angiosperms* (ed. B. M. Johri) Springer-Verlag, Berlin, pp. 123–57.

Braithwaite, A. F. (1964) A new type of apogamy in ferns. *New Phytol.*, **663**, 293–305.

Bremermann, H. J. (1980) Sex and polymorphism and strategies of host-pathogen interactions. *J. Theor. Biol.*, **87**, 641–702.

Brix, K. (1974) Sexual reproduction in *Eragrostis curvula* (Schrad.) Nees. *Z. Pflanzenzucht*, **71**, 25–32.

Brooks, L. D. (1988) The evolution of recombination rates, in *The Evolution of Sex* (eds R. E. Michod and B. R. Levin) Sinauer, Sunderland, MA, pp. 87–105.

Brown, W. V. and Emery, W. H. P. (1957) Apomixis in the Gramineae, tribe Andropogoneae: *Themeda triandra* and *Bothriochloa ischaemum*. *Bot. Gaz.*, **118**, 246–53.

Bryan, G. S. (1920) The fusion of ventral canal cell and egg in *Sphagnum subsecundum*. Am. J. Bot., **7**, 223–30.

Bryan, G. S. and Evans, R. I. (1956) Chromatin behaviour in the development and maturation of the egg nucleus of *Zamia umbrosa*. *Am. J. Bot.*, **43**, 610–46.

Burnett, J. H. (1975) *Mycogenetics*. John Wiley, London.

Burt, A. and Bell, G. (1987) Mammalian chiasma frequencies as a test of two theories of recombination. *Nature*, **326**, 803–5.

Burton, G. W. and Forbes, I. Jr (1960) The genetics and manipulation of obligate apomixis in common Bahia grass (*Paspalum notatum* Flugge). *Int. Grassl. Congr. Proc.*, **8**, 66–71.

Carson, H. L. (1961) Rare parthenogenesis in *Drosophila robusta*. *Am. Nat.*, **95**, 81–6.

Carson, H. L. (1967) Selection for parthenogenesis in *Drosophila mercatorium*. *Genetics*, **55**, 151–71.

Casselton, L. A. and Lewis, D. (1967) Dilution of gene products in the cytoplasm of heterokaryons in *Coprinus lagopus*. *Genet. Res.*, **9**, 63–71.

Catling, P. M. (1982) Breeding systems of Northeastern North American *Spiranthes* (Orchidaceae). *Can. J. Bot.*, **60**, 3017–39.

Chadefaud, M. (1941) Un probleme de botanique classique mais non resolu: celui des antheridies et des archegones des Phanerogames et plus particulierement des Angiosperms. *Rev. Sci.*, **79**, 479–82.

Chamberlain, J. S. (1935) *Gymnosperms*. University of Chicago Press, Chicago, reprinted in 1957 by the Johnson Reprint Corporation, New York.

Chao, C-Y (1974) Megasporogenesis and megagametogenesis in *Paspalum commersonii* and *P. longifolium* at two polyploid levels. *Bot. Not.*, **127**, 267–75.

Chao, C-Y (1980) Autonomous development of embryo in *Paspalum conjugatum* Berg. *Bot. Not.*, **133**, 215–22.

Charlesworth, B. (1980) The cost of sex in relation to mating system. *J. Theor. Biol.*, **84**, 655–71.

Charlesworth, B. and Charlesworth, D. (1985a) Genetic variation in recombination in *Drosophila*. I. Response to selection and preliminary genetic analysis. *Heredity*, **54**, 71–83.

Charlesworth, B. and Charlesworth, D. (1985b) Genetic variation in recombination in *Drosophila*. II. Genetic analysis of a high recombination stock. *Heredity*, **54**, 85–98.

Charnov, E. L. (1982) *The Theory of Sex Allocation*. Princeton University Press, Princeton.

Chatterji, A. K. and Timothy, D. H. (1969) Microsporogenesis and embryogenesis in *Pennisetum flaccidum* Griseb. *Crop. Sci.*, **9** 219–22.

Christensen, B. (1961) Studies on cyto-taxonomy and reproduction in the Enchytraeidae. *Hereditas*, **47**, 387–450.

Cohen, J. (1977) *Reproduction*. Butterworths, London.

Connor, H. E. (1979) Breeding systems in the grasses: a survey. *NZ J. Bot.*, **17**, 547–74.

Cooper, D. C. (1943) Haploid-diploid twin embryos in *Lilium* and *Nicotiana*. *Am. J. Bot.*, **30**, 408–13.

Courtice, G. R. M., Ashton, N. W. and Cove, D. J. (1978) Evidence for the restricted passage of metabolites into the sporophyte of the moss *Physcomitrella patens* (Hedw.) Br. *Eur. J. Bryol.*, **10**, 191–8.

Cousens, M. I. (1975) Gametophyte sex expression in some species of *Dryopteris. Am. Fern J.*, **65**, 39–42.

Cousens, M. I. (1988) Reproductive strategies of Pteridophytes. in *Plant Reproductive Ecology* (eds J. Lovett Doust and L. Lovett Doust), Oxford University Press, Oxford, pp. 307–28.

Crane, P. R. (1985) Phylogenetic analysis of seed plants and the origin of angiosperms. *Ann. Missouri. Bot. Gard.*, **72**, 716–93.

Crane, P. R. (1988) Major clades and relationships in the 'higher' gymnosperms, in *Origin and Evolution of Gymnosperms* (ed. C. B. Beck) Columbia University Press, New York, pp. 218–72.

Crew, F. A. E. (1965) *Sex Determination.* Methuen, London.

Cuellar, O. (1974) On the origin of parthenogenesis: the cytogenetic factors. *Am. Nat.*, **108**, 625–48.

Cuellar, O. (1987) The evolution of parthenogenesis: a historical perspective. in *Meiosis* (ed. P. B. Moens) Academic Press, Orlando, pp. 43–104.

Czapik, R. (1981) Embryology of *Rubus saxatilis* L. *Acta Biol. Crac. Ser. Bot.*, **23**, 7–13.

D'Amato, F. (1949) Triploidia e apomissia in *Statice oleaefolia* Scop. var. *confusa* Goodr. *Caryologia*, **2**, 71–84.

Darlington, C. D. (1939) *The Evolution of Genetic Systems*, Cambridge University Press, Cambridge.

Darwin, C. (1862) in *The Collected Papers of Charles Darwin* (ed. P. H. Barrett) University of Chicago Press, Chicago, 1977.

Davidson, E. H. (1976) *Gene Activity in Early Development*, Academic Press, New York.

Davies, W. E. and Young, N. R. (1966) Self-fertility in *Trifolium fragiferum. Heredity*, **21**, 615–24.

Davis, D. G. (1969) Chromosome behaviour under the influence of claret-nondisjunction in *Drosophila melanogaster. Genetics*, **61**, 577–94.

Dawkins, R. (1982) *The Extended Phenotype.* Oxford University Press, Oxford.

D'Cruz, R. and Reddy, P. S. (1971) Inheritance of apomixis in *Dichanthium. Ind. J. Genet. Plant Breed*, **31**, 451–60.

Dewees, A. A. (1970) Two-way selection for recombination rates in *Tribolium castaneum* (Abstract). *Genetics*, **64**, 516–17.

Dogra, P. D. (1966) Observations on *Abies pindrow* with a discussion on the question of occurrence of apomixis in Gymnosperms. *Silvae Genet.*, **15**, 11–20.

Döpp, W. (1932) Die apogamy bei *Aspidium remotum*. *Planta*, **17**, 86–152.

Dougherty, E. C. (1955) Comparative evolution and the origin of sexuality. *Syst. Zool.*, **4**, 145–69.

Doyle, J. A. and Donoghue, M. J. (1987) The origin of angiosperms: a cladistic approach, in *The Origins of Angiosperms and their Biological Consequences* (eds E. M. Friis, W. G. Chaloner and P. R. Crane) Cambridge University Press, Cambridge, pp. 17–49.

Drebes, G. (1977) Sexuality, in *The Biology of Diatoms* (ed. D. Werner) Blackwell, Oxford, pp. 250–83.

Duckett, J. G. and Renzaglia, K. S. (1988) Cell and molecular biology of bryophytes: ultimate limits to the resolution of phylogenetic problems. *Bot. J. Linn. Soc.*, **98**, 225–46.

Eames, A. J. (1955) The seed and *Ginkgo*. *J. Arnold Arb.*, **36**, 165–70.

Eenink, A. H. (1974) Matromorphy in *Brassica oleracea* L. V. Studies on quantitative characters of matromorphic plants and their progeny. *Euphytica*, **23**, 725–36.

Ellerstrom, S. and Zagorcheva, L. (1977) Sterility and apomictic embryo-sac formation in *Raphanobrassica*. *Hereditas*, **87**, 107–20.

Ellstrand, N. C. and Antonovics, J. (1985) Experimental studies of the evolutionary significance of sexual reproduction. II. A test of the density-dependent selection hypothesis. *Evolution*, **39**, 657–66.

Emsweller, S. L. and Jones, H. A. (1938) Crossing-over, fragmentation and formation of new chromosomes in an *Allium* species hybrid. *Bot. Gaz.*, **99**, 729–72.

Ernst, A. (1918) *Bastardierung als Ursache der Apogamie im Pflanzenreich*. Fischer, Jena.

Ernst, A. (1953) 'Basic numbers' und polyploidy und ihre Bedeutung fur des Heterostylie-problem. *Arch. Klaus Stift. Verebf.*, **28**, 1–159.

Ernst, A. and Bernard, Ch. (1912) Entwicklungsgeschichte des embrosackes, des Embryos und des Endosperms von *Burmannia coelestis* Don. *Ann. Jard. Bot. Buitenzorg II*, **11**, 234–57.

Esau, K. (1946) Morphology of reproduction in guayule and certain other species of *Parthenium*. *Hilgardia*, **17**, 61–101.

Evans, A. M. (1969) Interspecific relationships in the *Polypodium pectinatum – plumula* complex. *Ann. Miss. Bot. Gdn*, **55**, 193–293.

Evans, L. T. and Knox, R. B. (1969) Environmental control of reproduction in *Themeda australis*. *Aust. J. Bot.*, **17**, 375–89.

Fagerlind, F. (1946) Sporogenesis. Enbryosackentwicklung und pseudogame Samenbildung bei *Rudbeckia laciniata* L. *Acta Horti, Bergianai*, **14**, 39–90.

Fagerlind, F. (1947) Macrogametophyte formation in two agamospermous *Erigeron* species. *Acta Horti. Bergiani*, **14**, 221–47.

Falconer, D. S. (1981) *Introduction to Quantitative Genetics*, 2nd edn, Longman, London.

Favre-Duchartre, M. (1984) Homologies and phylogeny, in *Embryology of Angiosperms* (ed. B. M. Johri) Springer-Verlag, Berlin, pp. 697–734.

Felsenstein, J. (1965) The effect of linkage on directional selection. *Genetics*, **52**, 349–63.

Felsenstein, J. (1974) The evolutionary advantage of recombination. *Genetics*, **78**, 737–56.

Felsenstein, J. (1985) Recombination and sex: is Maynard Smith necessary? in *Evolution: Essays in Honour of John Maynard Smith* (eds P. J. Greenwood, P. H. Harvey and M. Slatkin) Cambridge University Press, Cambridge, pp. 209–20.

Felsenstein, J. (1988) Sex and the evolution of recombination, in *The Evolution of Sex* (eds R. E. Michod and B. R. Levin) Sinauer, Sunderland, MA, pp. 74–86.

Felsenstein, J. and Yokoyama, S. (1976) The evolutionary advantage of recombination. II. Individual selection for recombination. *Genetics*, **83**, 845–59.

Fincham, J. R. S. (1983) *Genetics*, Wright, Bristol.

Fincham, J. R. S., Day, P. R. and Radford, A. (1979) *Fungal Genetics*, 4th edn, Blackwell, Oxford.

Fisher, R. A. (1930) *The Genetical Theory of Natural Selection*, Clarendon press, Oxford.

Ford, H. and Richards, A. J. (1985) Isozyme variation within and between *Taraxacum* agamospecies in a single locality. *Heredity*, **55**, 289–91.

Foster, A. S. and Gifford, E. M. (1959) *Comparative Morphology of Vascular Plants*. W. H. Freeman, San Francisco.

Fritsch, F. E. (1965) *The Structure and Reproduction of the Algae*, Vol. 1. Cambridge University Press, Cambridge.

Fürnkranz, D. (1960) Cytogenetische Untersuchungen an *Taraxacum* im Raume von Wien. *Osterr. Bot. Zeitschr.*, **107**, 310–50.

Fürnkranz, D. (1961) Cytogenetische Untersuchungen an *Taraxacum* im Raume von Wien II. Hybriden zwischen *T. officinale* und *T. palustre. Osterr. Bot. Zeitschr.*, **108**, 408–15.

Fürnkranz, D. (1966) Untersuchungen an Populatione des *taraxacum* officinale-Komplexes in Kontaktgebiet der diploiden und polyploiden Biotypen. *Osterr. Bot. Zeitschr.*, **113**, 427–47.

Gale, M. D. and Rees, H. (1970) Genes controlling chiasma frequency in *Hordeum. Heredity*, **25**, 393–410.

Gastonby, G. L. and Haufler, C. H. (1976) Chromosome numbers and apomixis in the fern genus *Bommeria* (Gymnogrammaceae). *Biotropica*, **8**, 1–11.

Gates, R. R. and Goodwin, K. M. (1930) A new haploid *Oenothera*, with some considerations on haploidy in plants and animals. *J. Genet.*, **23**, 123–56.

Gentcheff, G. and Gustafsson, A. (1940) The balance system of meiosis in *Hieracium*. *Hereditas*, **26**, 209–49.

Gerstel, D. U., Hammond, B. L. and Kidd, C. (1953) An additional note on the inheritance of apomixis in guayule. *Bot. Gaz.*, **115**, 89–93.

Ghiselin, M. T. (1974) The economy of nature and the evolution of sex. University of California Press, Berkley, CA.

Giles, N. (1940) Spontaneous chromosome aberrations in *Tradescantia*. *Genetics*, **25**, 69–87.

Glesener, R. R. and Tilman, D. (1978) Sexuality and the components of environmental uncertainty: clues from geographic parthenogenesis in terrestrial animals. *Am. Nat.*, **112**, 659–73.

Grafen, A. (1985) A geometric view of relatedness. *Oxford Surv. Evol. Biol.*, , **2**, 28–89.

Graham, C. F. (1974) The production of parthenogenetic mammalian embryos and their use in biological research. *Biol. Rev.*, **49**, 399–422.

Grant, V. (1975) *Genetics of Flowering Plants*. Columbia University Press, New York.

Grant, V. (1979) *Plant Speciation*. Columbia University Press, New York.

Grant, V. (1981) *Plant Speciation*. 2nd. edn, Columbia University Press, New York.

Grime, J. P. (1979) *Plant Strategies and Vegetation Processes*. John Wiley and Son, Chichester, UK.

Gupta, P. K., Roy, R. P. and Singh, A. P. (1969) Aposporous apomicts: Seasonal variation in tetraploid *Dichanthium annulatum*. *Port Acta Biol. Ser. A*, **11**, 253–60.

Gustafsson, A. (1936) Studies on the mechanism of parthenogenesis. *Hereditas*, **21**, 1–112.

Gustafsson, A. (1946–47) Apomixis in higher plants. I-III. *Lunds Univ. Arsskrift*, **42**, 1–67; **43**, 69–179; **43**, 181–371.

Gustafsson, A. (1947) Mutations in agricultural plants. *Hereditas*, **33**, 1–100.

Gustafsson, A. (1948) Polyploidy, life form and vegetative reproduction. *Hereditas*, **34**, 1–22.

Haig, D. and Westoby, M. (1988) A model for the evolution of heterospory. *J. Theor. Biol.*, **134**, 257–72.

Haig, D. and Westoby, M. (1989) Selective forces in the emergence of the seed habit. *Biol. J. Linn. Soc.*, **38**, 215–38.

Haigh, D. (1986) Conflicts among megaspores. *J. Theor. Biol.*, **123**, 471–80.

Hair, J. B. (1956) Subsexual reproduction in *Agropyron*. *Heredity*, **10**, 129–60.

Hakansson, A. (1943) Die Entwicklung des Embryosacks und die Befruchtung bei *Poa alpina*. *Hereditas*, **29**, 25–61.

Hakansson, A. (1951) Parthenogenesis in *Allium*. *Bot. Not.*, **104**, 143–79.

Hakansson, A. and Levan, A. (1957) Endo-duplicational meiosis in *Allium odorum*. *Hereditas*, **43**, 179–200.

Halac, N. de and Harte, C. (1975) Female gametophyte competence in relation to polarisation phenomena during the megagametogenesis and development of the embryo sac in the genus *Oenothera*, in *Gamete Competition in Plants and Animals* (ed. D. L. Mulcahy) North-Holland, Amsterdam, pp. 43–56.

Hamilton, W. D. (1964a) The genetical evolution of social behaviour. I. *J. Theor. Biol.*, **7**, 1–16.

Hamilton, W. D. (1964b) The genetical evolution of social behaviour. II. *J. Theor. Biol.*, **7**, 17–52.

Hamilton, W. D. (1970) Selfish and spiteful behaviour in an evolutionary model. *Nature*, **228**, 1218–20.

Hamilton, W. D. (1980) Sex versus non-sex versus parasite. *Oikos*, **35**, 282–90.

Hamilton, W. D. (1982) Pathogens as causes of genetic diversity in their host populations. in *Population Biology of Infectious Diseases* (eds R. M. Anderson and R. M. May) Springer-Verlag, New York, pp. 269–96.

Hamilton, W. D., Axelrod, R. and Tanese, R. (1990) Sexual reproduction as an adaptation to resist parasites (A review). *Proc. Natl. Acad. Sci. USA*, **87**, 3566–73.

Hanna, W. W., Powell, J. B., Millot, J. C. and Burton, G. W. (1973) Cytology of obligate sexual plants in *Panicum maximum* Jacq. and their use in controlled hybrids. *Crop Sci.*, **13**, 695–7.

Hardwick, R. C. (1986) Physiological consequences of modular growth in plants. *Philos. Trans. R. Soc. Lond. [Biol]*, **313**, 161–73.

Harlan, J. R., Brooks, M. H., Borgaonkar, D. S. and De Wet, J. M. J. (1964) Nature and inheritance of apomixis in *Bothriochloa* and *Dichanthium*. *Bot. Gaz.*, **125**, 41–6.

Harlan, J. R. and de Wet, J. M. J. (1975) On O. Winge and a prayer: the origins of polyploidy. *Bot. Rev.*, **41**, 361–90.

Harper, J. L. (1977) *Population Biology of Plants*, Academic Press, London.

Harper, J. L., Rosen, B. R. and White, J. (1986) Preface to 'The growth and form of modular organisms'. *Philos. Trans. R. Soc. Lond. [Biol]*, **313**, 3–5.

Haskell, G. (1953) Quantitative variation in subsexual *Rubus*. *Heredity*, **7**, 409–18.

Haskell, G. (1960) Role of the male parent in apomictic *Rubus* species. *Heredity*, **14**, 101–13.

Haskell, G. (1966) The history, taxonomy and breeding system of apomictic British Rubi, in *Reproductive Biology and Taxonomy of Vascular Plants* (ed. J. G. Hawkes) Pergamon, Oxford, pp. 141–51.

Hewitt, G. M. (1975) A new hypothesis for the origin of the partheno-genetic grasshopper *Moraba virgo*. *Heredity*, **34**, 117–23.

Heywood, V. H. (ed.) (1978) *Flowering Plants of the World*. Oxford University Press, Oxford.

Hill, W. G. and Robertson, A. (1966) The effect of linkage on limits to artificial selection. *Genet. Res.*, **8**, 269–94.

Hogarth, P. J. (1978) *Biology of Reproduction*, Blackie, Glasgow.

Holliday, R. (1984) The biological significance of meiosis. in *Controlling Events in Meiosis* (eds C. E. Evans and H. G. Dickenson) S. E. B. Symposia 38, Cambridge University Press, Cambridge.

Huxley, J. S. (1912) *The Individual in the Animal Kingdom*. Cambridge University Press, Cambridge.

Ingold, C. T. (1973) *The Biology of Fungi*, 3rd edn. Hutchinson, London.

Ivanov, M. A. (1938) Experimental production of haploids in *Nicotiana rustica* L. *Genetica*, **20**, 295–397.

Jackson, I. J. (1989) The mouse. in *Genes and Embryos* (eds D. M. Glover and B. D. Hames) IRL Press and Oxford University Press, Oxford, pp. 165–221.

Jaenike, J. (1978) An hypothesis to account for the maintenance of sex within populations. *Evol. Theor.*, **3**, 191–4.

Janzen, D. H. (1977) What are dandelions and aphids? *Am. Nat.*, **111**, 586–9.

Jeffreys, E. C. (1895) Polyembryony in *Erythronium americanum*. *Ann. Bot.*, **9**, 537–41.

John, B. S. (1977) *The Ice Age, Past and Present*. Collins, London.

Johri, B. M. (ed.) (1984) *Embryology of Angiosperms*. Springer-Verlag, Berlin.

Johri, B. M. and Bhatnagar, S. P. (1955) A contribution to the morphology and life history of *Aristolochia*. *Phytomorphology*, **5**, 123–37.

Kapil, R. N. and Bhatnagar, A. K. (1981) Ultrastructure and biology of female gametophyte in flowering plants. *Int. Rev. Cytol.*, **70**, 291–341.

Kehr, A. E. (1951) Monoploidy in *Nicotiana. J. Hered.*, **42**, 107–12.

Khan, R. (1943) Contributions to the morphology of *Ephedra foliata* Boiss. II. Fertilization and embryogeny. *Proc. Natl. Acad. Sci., India*, **13**, 357–75.

Khokhlov, S. S. (1976) *Apomixis and Breeding*. Amerind, New Delhi.

Kimber, G. and Riley, R. (1963) Haploid angiosperms. *Bot. Rev.*, **29**, 480–531.

King, L. M. and Schaal, B. A. (1990) Genotype variation within asexual lineages of *Taraxacum officinale. Proc. Natl. Acad. Sci. USA*, **87**, 998–1002.

King, R. L. and Slifer, E. H. (1934) Insect development. VIII. Maturation and early development of unfertilised grasshopper eggs. *J. Morphol.*, **56**, 603–20.

Klekowski, E. J. Jr (1969) Reproductive biology of the pteridophyta. II. Theoretical considerations. *Bot. J. Linn. Soc.*, **62**, 347–59.

Klekowski, E. J. Jr (1970) Population and genetic studies of a homosporous fern – *Osmunda regalis. Am. J. Bot.*, **57**, 1122–38.

Klekowski, E. J. Jr (1973a) Genetic load in *Osmunda regalis* populations. *Am. J. Bot.*, **60**, 146–54.

Klekowski, E. J. Jr (1973b) Sexual and subsexual systems in homosporous pteridophytes: a new hypothesis. *Am. J. Bot.*, **60**, 535–44.

Klekowski, E. J. Jr (1979) The genetics and reproductive biology of ferns. in *The Experimental Biology of Ferns* (ed. A. F. Dyer) Academic Press, London, pp. 133–70.

Knox, R. B. (1967) Apomixis: seasonal and population differences in a grass. *Science*, **157**, 325–6.

Knox, R. B. and Heslop-Harrison, J. (1963) Experimental control of aposporous apomixis in a grass of the Andropogonea. *Bot. Not.*, **116**, 127–41.

Kostoff, D. (1942) The problem of haploidy. (Cytogenetic studies in *Nicotiana* hybrids and their bearing on some other cytogenetic problems.) *Bib. Genet.*, **13**, 1–148.

Kryolova, V. V. (1976) Apospory and polyembryony in apple. in *Apomixis and Breeding* (ed. S. S. Khokslov) Amerind, New Delhi, pp. 124–9.

Lakshmanan, K. K. and Ambegaokar, K. B. (1984) Polyembryony. in *Embryology of Angiosperms* (ed. B. M. Johri) Springer-Verlag, Berlin, pp. 445–74.

Lal, M. (1984) The culture of bryophytes including apogamy, apospory, parthenogenesis and protoplasts, in *The Experimental Biology of*

Bryophytes (eds A. F. Dyer and J. G. Duckett) Academic Press, London, pp. 97–115.

Lamb, R. Y. and Willey, R. B. (1979) Are parthenogenetic and related bisexual insects equal in fertility? *Evolution*, **33**, 774–5.

Lane, D. M. (1985) A quantitative study of the mosses of eastern North America. *Monog. Syst. Bot. Missouri Bot. Gard*, **11**, 45–50.

Levin, D. A. (1975) Pest pressure and recombination systems in plants. *Am. Nat.*, **109**, 437–51.

Levin, D. A. (1983) Polyploidy and novelty in flowering plants. *Am. Nat.*, **122**, 1–25.

Lewin, B. (1980) *Gene Expression, 2. Eucaryotic Chromosomes* 2nd edn, John Wiley, New York.

Lewis, W. H. (ed.) (1980) *Polyploidy: Biological Relevance*. Plenum Press, New York.

Lewis, W. M. Jr (1983) Interruption of synthesis as a cost of sex in small organisms. *Am. Nat.*, **121**, 825–34.

Lewis, W. M. Jr (1987) The cost of sex, in *The Evolution of Sex and Its Consequences* (ed. S. C. Stearns) Birkhauser Verlag, Basel, pp. 33–57.

Liljefors, A. (1955) Cytological studies in *Sorbus. Acta Horti. Bergiani.*, **16**, 47–113.

Lloyd, D. G. (1980) Demographic factors and mating patterns in Angiosperms, in *Demography and Evolution in Plant Populations* (ed. O. T. Solbrig) Blackwell Scientific, Oxford, pp. 67–88.

Lloyd, R. M. (1974) Reproductive biology and evolution in the pteridophytes. *Ann. Mo. Bot. Gdn.*, **61**, 318–31.

Lobban, C. S. and Wynne, M. J. (eds) (1981) *The Biology of Seaweeds.* Blackwell, Oxford

Lokki, J. (1976a) Genetic polymorphism and evolution in parthenogenetic animals. VII. The amount of heterozygosity in diploid populations. *Hereditas*, **83**, 57–64.

Lokki, J. (1976b) Genetic polymorphism and evolution in parthenogenetic animals. VIII. Heterozygosity in relation to polyploidy. *Hereditas*, **83**, 65–72.

Longton, R. E. and Miles, C. J. (1982) Studies on the reproductive biology of mosses. *J. Hattori Bot. Lab.*, **52**, 219–40.

Love, A., Love, D. and Pichi Sermolli, R. E. G. (1977) *Cytotaxonomical Atlas of the Pteridophyta*, Cramer, Vaduz.

Lovett Doust, J. and Lovett Doust, L. (eds) (1988) *Plant Reproductive Ecology*, Oxford University Press, Oxford.

Lovis, J. D. (1977) Evolutionary patterns and processes in ferns. *Adv. Bot. Res.*, **4**, 229–415.

Lynch, M. (1984) Destabilizing hybridisation, general-purpose genotypes and geographic parthenogenesis. *Q. Rev. Biol.*, **59**, 257–90.

Malecka, J. (1962) Cytological studies in the genus *Taraxacum*. *Acta Biol. Crac. Ser. Bot.*, **5**, 117–36.

Malecka, J. (1965) Embryological studies in *Taraxacum palustre*. *Acta Biol. Crac. Ser. Bot.*, **8**, 223–35.

Malecka, J. (1967a) Chromosome numbers of 5 *Taraxacum* species from Mongolia. *Acta Biol. Crac. Ser. Bot.*, **10**, 73–83.

Malecka, J. (1967b) Cyto-embryological studies in *Taraxacum scanicum* Dt. *Acta Biol. Crac. Ser. Bot.*, **10**, 195–206.

Malecka, J. (1969) Further cyto-taxonomic studies in the genus *Taraxacum* section *Erythrosperma*. Dt. I. *Acta Biol. Crac. Ser. Bot.*, **12**, 57–69.

Malecka, J. (1970) Cyto-taxonomic studies in the genus *Taraxacum* section *Palustria* Dahlst. *Acta Biol. Crac. Ser. Bot.*, **13**, 155–68.

Malecka, J. (1971) Cyto-taxonomical and embryological investigations on a natural hybrid between *Taraxacum kok-saghyz* and *T. officinale* Web. and their putative parent species. *Acta Biol. Crac. Ser. Bot.*, **14**, 179–97.

Malecka, J. (1972) Further cyto-taxonomic studies in the genus *Taraxacum* section *Palustria* Dahlst. *Acta Biol. Crac. Ser. Bot.*, **15**, 113–26.

Manning, J. T. and Dickson, D. P. E. (1986) Asexual reproduction, polyploidy and optimal mutation rates. *J. Theor. Biol.*, **118**, 485–9.

Manton, I. (1950) *Problems of Cytology and Evolution in the Pteridophyta*, Cambridge University Press, Cambridge.

Marklund, G. and Rousi, A. (1961) Outlines of evolution in the pseudogamous *Ranunculus auricomus* group in Finland. *Evolution*, **15**, 510–22.

Mathew, C. J. (1980) Embryological studies in Hamamelidaceae: Development of female gametophyte and embryogeny in *Hamamelis virginiana*. *Phytomorphology*, **30**, 172–80.

Mathew, C. J. and Chaphekar, M. (1977) Development of female gametophyte and embryogeny in *Stachyurus chinensis*. *Phytomorphology*, **27**, 68–78.

Maynard Smith, J. (1971) The origin and maintenance of sex, in *Group Selection* (ed. G. Williams) Aldine-Atherton, Chicago, pp. 163–75.

Maynard Smith, J. (1976) A short term advantage for sex and recombination through sib-competition. *J. Theor. Biol.*, **63**, 245–58.

Maynard Smith, J. (1978) *The Evolution of Sex*, Cambridge University Press, Cambridge.

Maynard Smith, J. (1980) Selection for recombination in a polygenic model. *Genet. Res.*, **35**, 269–77.

Maynard Smith, J. (1988a) The evolution of recombination. in *The Evolution of Sex* (eds R. E. Michod and B. R. Levin) Sinauer, Sunderland, MA, pp. 106–25.

Maynard Smith, J. (1988b) Selection for recombination in a polygenic model – the mechanism. *Genet. Res.*, **51**, 59–63.

Maynard Smith, J. (1989) *Evolutionary Genetics*, Oxford University Press, Oxford.

Mayr, E. (1970) *Populations, Species, and Evolution*, Belknap, Cambridge, MA.

McCauley, D. E., Whittier, D. P. and Reilly, L. M. (1985) Inbreeding and the rate of self-fertilization in a grape fern, *Botrychium dissectum*. *Am. J. Bot.*, **72**, 1978–81.

McGrath, J. and Solter, D. (1984) Completion of mouse embryogenesis requires both the maternal and paternal genomes. *Cell*, **37**, 179–83.

McGrath, J. and Solter, D. (1986) Nucleocytoplasmic interactions in the mouse embryo. *J. Embryol. Morphol.*, **97** (Suppl), 277–89.

Mehra, P. N. and Sanhu, R. S. (1976) Morphology of the fern *Anogramma leptophylla*. *Phytomorphology*, **26**, 60–76.

Michaels, H. J. and Bazzaz, F. A. (1986) Resource allocation and demography of sexual and apomictic *Antennaria parlinii*. *Ecology*, **67**, 27–36.

Michod, R. E. and Levin, B. R. (eds) (1988) *The Evolution of Sex*, Sinauer, Sunderland, MA.

Miles, C. J. and Longton, R. E. (1990) The role of spores in reproduction in mosses. *Bot. J. Linn. Soc.*, **104**, 149–73.

Mishler, B. D. (1988) Reproductive ecology of bryphytes, in *Plant Reproductive Ecology* (eds J. L. Doust and L. L. Doust) Oxford University Press, Oxford, pp. 285–306.

Mogensen, G. S. (1983) The spore, in *New Manual of Bryology* (ed. R. M. Schuster) Hattori Botanical Laboratory, Japan, pp. 325–42.

Mogie, M. (1982) *The Status of Taraxacum Agamospecies*. Unpublished PhD thesis, University of Newcastle.

Mogie, M. (1985) Morphological, developmental and electrophoretic variation within and between obligately apomictic *Taraxacum* species. *Biol. J. Linn. Soc.*, **24**, 207–16.

Mogie, M. (1986a) Automixis: its distribution and status. *Biol. J. Linn. Soc.*, **28**, 321–9.

Mogie, M. (1986b) On the relationship between asexual reproduction and polyploidy. *J. Theor. Biol.*, **122**, 493–8.

Mogie, M. (1988) A model for the evolution and control of generative apomixis. *Biol. J. Linn. Soc.*, **35**, 127–53.

Mogie, M. (1990) Homospory and the cost of asexual reproduction. *Evolution*, **44**, 1707–10.

Mogie, M. and Ford, H. (1988) Sexual and asexual *Taraxacum* species. *Biol. J. Linn. Soc.*, **35**, 155–68.

Mogie, M. and Richards, A. J. (1983) Satellited chromosomes, systematics and phylogeny in *Taraxacum* (Asteraceae). *Plant Syst. Evol.*, **141**, 219–29.

Mogie, M. and Hutchings, M. J. (1990) Phylogeny, ontogeny and clonal growth in vascular plants, in *Clonal Growth in Plants: Regulation and Function* (eds J. van Groenendael and H. de Kroon) SPB Publishing, The Hague, The Netherlands, pp. 3–22.

Moran, G. F., Bell, J. C. and Hilliker, A. J. (1983) Greater meiotic recombination in male vs female gametes in *Pinus radiata. J. Hered.*, **74**, 62.

Morita, T., Menken, S. B. J. and Sterk, A. A. (1990) Hybridization between European and Asian dandelions (*Taraxacum* section *Ruderalia* and section *Mongolica*). *New. Phytol.*, **114**, 519–29.

Muller, H. J. (1932) Some genetic aspects of sex. *Am. Nat.*, **66**, 118–38.

Muller, H. J. (1964) The relation of recombination to mutational advance. *Mutat. Res.*, **1**, 2–9.

Muntzing, A. (1934) Chromosome fragmentation in a *Crepis* hybrid. *Hereditas*, **19**, 284–302.

Muntzing, A. (1940) Further studies on apomixis and sexuality in *Poa. Hereditas*, **26**, 115–90.

Muntzing, A. and Muntzing, G. (1945) The mode of reproduction of hybrids between sexual and apomictic *Potentilla argentea. Bot. Not.*, **98**, 49–71.

Naf, U. (1979) Antheridiogens and antheridial development, in *The Experimental Biology of Ferns* (ed. A. F. Dyer) Academic Press, London, pp. 435–70.

Nath, J., Swaminathan, M. S. and Mehra, K. L. (1970) Cytological studies in the Tribe Paniceae, Gramineae. *Cytologia*, **35**, 111–31.

Newton, M. E. (1984) The cytogenetics of bryophytes, in *The Experimental Biology of Bryophytes* (eds A. F. Dyer and J. G. Duckett) Academic Press, London, pp. 65–96.

Newton, M. E. (1990) Genetic structure of hepatic species. *Bot. J. Linn. Soc.*, **104**, 215–29.

Nijs, J. C. M. den and Sterk, A. A. (1980) Cytogeographical studies of *Taraxacum* sect. *Taraxacum* (= sect. *Vulgaria*) in Central Europe. *Bot. Jahrb. Syst.*, **101**, 527–54.

Nogler, G. A. (1971) Genetik der Aposporie bei *Ranunculus auricomus* s. 1. W. Kock. I. Embryologie. *Ber Schweiz Bot Ges.*, **81**, 139–79.

Nogler, G. A. (1984) Gametophytic apomixis, in *Embrology of Angiosperms* (ed. B. M. Johri) Springer-Verlag, Berlin, pp. 475–518.

Nokkala, S. and Nokkala, C. (1983) Achiasmatic male meiosis in two species of *Saluda* (Saldidae, Hemiptera). *Hereditas*, **99**, 131–4.

Nokkala, S. and Nokkala, C. (1984) Achiasmatic male meiosis in the heteropteran genus *Nabis* (Nabidae, Hemiptera). *Hereditas*, **101**, 31–5.

Nygren, A. (1946) The genesis of some Scandinavian species of *Calamagrostis*. *Hereditas*, **32**, 131–262.

Nygren, A. (1954) Apomixis in the angiosperms. *Bot. Rev.*, **20**, 577–649.

Nygren, A. (1967) Apomixis in the angiosperms. *Encyclopedia of Plant Physiol.*, **18**, 551–96.

Ohno, S. (1970) *Evolution by Gene Duplication*, Springer-Verlag, New York.

Okabe, S. (1932) Parthenogenesis bei *Ixeris dentata*. *Bot. Mag. (Tokyo)*, **46**, 518–23.

Øllgaard, H. (1983) *Hamata*: a new section of *Taraxacum*. *Plant Syst. Evol.*, **141**, 199–217.

Orr-Ewing, A. L. (1957) Possible occurrence of viable unfertilized seeds in Douglas-fir. *Forest Sci.*, **3**, 243–8.

Ostenfeld, C. H. (1910) Further studies on the apogamy and hybridization of the Hieracia. *Z. Indukt. Abstamm Vererbungslehre*, **3**, 241–85.

Ostenfeld, C. H. (1912) Experiments on the origin of species in the genus *Hieracium* (apogamy and hybridism). *New Phytol*, **11**, 347–53.

Page, C. N. (1979) The diversity of ferns. An ecological perspective, in *The Experimental Biology of Ferns* (ed. A. F. Dyer) Academic Press, London, pp. 9–56.

Pandey, K. K. (1955) Seed development in diploid, tetraploid and diploid-tetraploid crosses of *Trifolium pratense*. *Ind. J. Genet. Plant Breeding*, **15**, 25–35.

Pant, D. D. and Singh, R. (1989) On the possible occurrence of anisospory in some Hepaticae. *Bot. J. Linn. Soc.*, **100**, 183–96.

Parihar, N. S. (1965) *An Introduction to Embryophyta. I. Bryopyta*, Central Book Depot, Allahabad.

Parker, G. A. (1979) An evolutionarily stable strategy approach to indiscriminate spite. *Nature*, **279**, 419–21.

Peacock, A. D. and Harrison, J. W. H. (1926) Hybridity, parthenogenesis and segregation. *Nature*, **117**, 378–9.

Pechan, P. M. (1988) Ovule fertilization and seed number per pod determination in oil seed rape (*Brassica napus*), *Ann. Bot.*, **61**, 201–7.

Pijnacker, L. P. (1969) Automictic parthenogenesis in the stick insect *Bacillus rossius* Rossi (Chelentoptera, Phasmidae). *Genetica*, **40**, 393–9.

Pijnacker, L. P. and Ferwerda, M. A. (1976) Experiments on blocking and unblocking of first meiotic metaphase in eggs of the parthenogenetic stick insect *Carausius morosus* Br. (Phasmida, Insecta). *J. Embryol. Exp. Morphol.*, **36**, 383–94.

Porsild, A. E. (1965) The genus *Antennaria* in eastern Arctic and subarctic America. *Bot. Tidskr.*, **61**, 22–55.

Powers, L. (1945) Fertilization without reduction in guayule (*Parthenium argentatum* Gray) and a hypothesis as to the evolution of apomixis and polyploidy. *Genetics*, **30**, 323–46.

Quarin, C. L. and Hanna, W. W. (1980) Effect of three ploidy levels on meiosis and mode of reproduction in *Paspalum hexastachyum*. *Crop Sci.*, **20**, 69–75.

Queller, D. C. (1983) Kin selection and conflict in seed maturation. *J. Theor. Biol.*, **100**, 153–72.

Raghavan, V. (1976) *Experimental Embryogenesis in Vascular Plants*. Academic Press, London.

Reddy, P. S. and D'Cruz, R. (1966) Androgenetic tetraploids in *Dichanthium* Willemet. *The Nucleus*, **9**, 167–72.

Reddy, P. S. and D'Cruz, R. (1969) Mechanism of apomixis in *Dichanthium annulatum* (Forssk) Stapf. *Bot. Gaz.*, **130**, 71–9.

Reese, W. D. (1984) Reproductivity, fertility, and range of *Syrrhopodon texanus* Sull. (Musci: Calymperaceae), a North American endemic. *Bryologist*, **87**, 217–22.

Reik, W., Collick, A., Norris, M. L., Barton, S. C. and Surani, M. A. H. (1987) Genomic imprinting determines methylation of parental alleles in transgenic mice. *Nature*, **328**, 248–51.

Rice, W. R. (1983) Sexual reproduction: an adaptation reducing parent-offspring contagion. *Evolution*, **37**, 1317–20.

Richards, A. J. (1970a) Hybridization in *Taraxacum. New Phytol.*, **69**, 1103–21.

Richards, A. J. (1970b) Eutriploid facultative agamospermy in *Taraxacum. New Phytol.*, **69**, 761–74.

Richards, A. J. (1972) The karyology of some *Taraxacum* species from alpine regions of Europe. *Bot. J. Linn. Soc.*, **65**, 47–59.

Richards, A. J. (1973) The origin of *Taraxacum* agamospecies. *Bot. J. Linn. Soc.*, **66**, 189–211.

Richards, A. J. (1986) *Plant Breeding Systems*. George Allen and Unwin, London.

Richards, A. J. (1989) A comparison of within-plant karyological heterogeneity between agamospermous and sexual *Taraxacum* (*Compositae*) as assessed by the nucleolar organiser chromosome. *Plant Syst. Evol.*, **163**, 177–85.

Richardson, D. H. S. (1981) *The Biology of Mosses*. Blackwell Scientific, Oxford.

Rollins, R. C. (1945) Evidence for genetic variation among apomictically produced plants of several F1 progenies of guayule (*Parthenium argentatum*) and mariola (*P. incanum*). *Am. J. Bot.*, **32**, 554–60.

Rollins, R. C. (1949) Sources of genetic variation in *Parthenium argentatum* Gray (Compositae). *Evolution*, **3**, 358–68.

Round, F. E. (1973) *The Biology of Algae*, 2nd edn, Edward Arnold, London.

Rousi, A. (1955) Cytological observations on the *Ranunculus auricomus* group. *Hereditas*, **41**, 516–18.

Runquist, E. W. (1968) Meiotic investigations in *Pinus sylvestris* L. *Hereditas*, **60**, 77–128.

Russell, S. D. (1979) Fine structure of megagametophyte development in *Zea mays*. *Can. J. Bot.*, **57**, 1093–110.

Rutishauser, A. (1946) Uber kreuzungsversuche mit pseudogamen *Potentillen*. Sechster Jahresber. Schweiz. Ges. Vererb. S.S.G. *Arch. d. Julius Klaus-Stiftung f. Ver., Soz. und Rass.*, **21**, 469–72.

Rutishauser, A. (1947) Untersuchungen uber die genetik der aposporie bei pseudogamen *Potentillen*. *Experientia*, **3**, 204.

Rychlewski, J. (1961) Cyto-embryological studies in the apomictic species *Nardus stricta* L. *Acta Biol. Crac. Ser. Bot.*, **4**, 1–23.

Sadasivaiah, R. S. and Weijer, J. (1981) Cytogenetics of some natural intergeneric hybrids between *Elymus* and *Agropyron* species. *Can. J. Genet. Cytol.*, **23**, 131–40.

Saran, S. and de Wet, J. M. J. (1970) The mode of reproduction in *Dichanthium intermedium* (Gramineae). *Bull. Torrey Bot. Club*, **97**, 6–13.

Saran, S. and de Wet, J. M. J. (1976) Environmental control of reproduction in *Dichanthium intermedium*. *J. Cytol. Genet.*, **11**, 22–8.

Savidan, Y. (1975) Heredite de l'apomixie. Contribution a l'etude de l'heredite de l'apomixie sur *Panicum maximum* Jacq. (analyse des sacs embryonnaires). *Cah ORSTOM Ser. Biol.*, **10**, 91–5.

Savidan, Y. (1980) Chromosomal and embryological analysis in sexual × apomictic hybrids of *Panicum maximum* Jacq. *Theor. Appl. Genet.*, **57**, 153–6.

Savidan, Y. (1981) Genetics and utilization of apomixis for the improvement of guinea grass (*Panicum maximum* Jacq.). *Proc. XIV Int. Brassl. Congr. Lexington Ky*, **1981**, 182–4.

Savidan, Y. (1982) Nature et heredite de l'apomixie chez *Panicum maximum* Jacq. *Trav. Doc. ORSTOM*, **153**, 1–159.

Schmid, B. (1990) Some ecological and evolutionary consequences of modular organization and clonal growth in plants. *Evol. Trends Plants*, **4**, 25–34.

Schmidt, H. (1964) Beitrage zur Zuchtung apomiktischer Apfelunterlagen I. Zytogenetische und embryologische Untersuchungen. *Z. Pflanzenzucht*, **52**, 27–102.

Schmitt, J. and Antonovics, J. (1986) Experimental studies of the evolutionary significance of sexual reproduction. IV. Effect of neighbour relatedness and aphid infestation on seedling performance. *Evolution*, **40**, 830–6.

Schuster, R. M. (1966) *The Hepaticae and Anthocerotae of North America East of the Hundredth Meridian*, Vol. 1, Columbia University Press, New York.

Sedgewick, P. J. (1924) Life history of *Encephalartos*. *Bot. Gaz.*, **77**, 300–10.

Senathirajah, S. and Lewis, D. (1975) Resistance to amino acid analogues in *Coprinus*: dominance modifier genes and dominance in dikaryons and diploids. *Genet. Res.*, **25**, 95–107.

Serrano, J. (1981) Male achiasmatic meiosis in *Caraboidea* (Coleoptera, Adephaga). *Genetica*, **57**, 131–7.

Shaw, D. D. (1972) Genetic and environmental components of chiasma control. II. The response to selection in *Schistocerca*. *Chromosoma*, **37**, 297–308.

Shaw, W. R. (1897) Parthenogenesis in *Marsilea*. *Bot. Gaz.*, **24**, 114–17.

Singh, D., Kaul, V. and Dathan, A. S. R. (1974) Cytological studies in the genus *Taraxacum* Weber. *Proc. Ind. Acad. Sci.*, **80**, 82–91.

Smith, A. J. E. (1978) Cytogenetics, biosystematics and evolution in Bryopyhta. *Adv. Bot. Res.*, **6**, 195–276.

Smith, G. L. (1963) Studies in *Potentilla* L. I. Embryological investigations into the mechanism of agamospermy in British *P. tabernaemontani* Aschers. *New Phytol.*, **62**, 264–82.

Smith, R. L. (1984) *Sperm Competition and the Evolution of Animal Mating Systems*. Academic Press, Orlando, FL.

Solter, D. (1987) Inertia of the embryonic genome in mammals. *Trends Genet.*, **3**, 23–7.

Soltis, D. E. and Soltis, P. S. (1986) Electrophoretic evidence for inbreeding in the fern *Botrychium virginianum* (Ophioglossaceae). *Am. J. Bot.*, **73**, 588–92.

Soltis, D. E. and Soltis, P. S. (1987) Polyploidy and breeding systems in homosporous pteridophytes: a reevaluation. *Am. Nat.*, **130**, 219–32.

Soreng, R. J. (1984) Dioecy and apomixis in the *Poa fendleriana* complex (Poaceae). *Am. J. Bot.*, **71**, 189.

Sørensen, T. (1958) Sexual chromosome-aberrants in triploid apomictic *Taraxaca*. *Bot. Tidsskrift*, **54**, 1–22.

Sørensen, T. and Gudjónsson, G. (1946) Spontaneous chromosome-aberrants in apomictic *Taraxaca*. *Biol. Skrift.*, **4**, 2–48.

Souciet, J. L. (1978) Controle de l'inhibition de la germination chez une espece apomictique, le *Panicum maximum*. These 2385, Univ. Paris-Sud.

Sparvoli, E. (1960) Osservazioni cito-embriologiche in *Eupatorium riparium* Reg. II. Megasporogenesi e sviluppo del gametofito femminile. *Ann. di Bot. (Rome)*, **26**, 481–504.

Sporne, K. R. (1965) *The Morphology of Gymnosperms*. Hutchinson University Library, London.

Sporne, K. R. (1975) *The Morphology of Pteridophytes*, 4th edn, Hutchinson, London.

Springer, E. (1935) Uber apogame (vegetativ enstandene) Sporogone an der bivalenten Rasse des Laubmooses *Phascum cuspidatum*. *Z. Induct. Abstamm.-Vererbungsl.*, **69**, 249–62.

Stalker, H. O. (1954) Parthenogenesis in *Drosophila*. *Genetics*, **39**, 4–34.

Stalker, H. D. (1956) On the evolution of parthenogenesis in *Lonchoptera* (Diptera). *Evolution*, **10**, 345–59.

Stearns, S. C. (ed.) (1987a) *The Evolution of Sex and Its Consequences*. Birkhauser Verlag, Basel.

Stearns, S. C. (1987b) Why sex evolved and the differences it makes, in *The Evolution of Sex and Its Consequences* (ed. S. C. Stearns) Birkhauser Verlag, Basel, pp. 15–31.

Stebbins, G. L. (1941) Apomixis in the angiosperms. *Bot. Rev.*, **7**, 507–42.

Stebbins, G. L. (1950) *Variation and Evolution in Plants*, Columbia University Press, New York.

Stebbins, G. L. (1971) *Chromosomal Evolution in Plants*, Edward Arnold, London.

Stebbins, G. L. and Jenkins, J. A. (1939) Aposporic development in North American species of *Crepis*. *Genetica*, **21**, 191–224.

Stewart, T. A., Hecht, N. B., Hollingshead, P. G., Johnson, P. A., Leong, J. A. C. and Pitts, S. L. (1988) Haploid-specific transcription of protamine-*myc* and protamine-T-antigen fusion genes in transgenic mice. *Mol. Cell. Biol.*, **8**, 1748–55.

Stewart, W. N. (1983) *Paleobotany and the Evolution of Plants*, Cambridge University Press, Cambridge.

Strobeck, C. (1975) Selection in a fine-grained environment. *Am. Nat.*, **109**, 419–25.

Sturtevant, A. H. (1939) High mutation frequency induced by hybridisation. *Proc. Natl. Acad. Sci. USA*, **25**, 308–10.

Sturtevant, A. H. and Mather, K. (1938) The interrelations of inversions, heterosis and recombination. *Am. Nat.*, **72**, 447–52.

Sullivan, V. I. (1976) Diploidy, polyploidy, and agamospermy among species of *Eupatorium* (Compositae). *Can. J. Bot.*, **54**, 2907–17.

Suomalainen, E. and Saura, A. (1973) Genetic polymorphism and evolution in parthenogenetic animals. I. Polyploid Curculionidae. *Genetics*, **74**, 489–508.

Suomalainen, E., Saura, A. and Lokki, J. (1976) Evolution of parthenogenetic insects. *Evol. Biol.*, **9**, 209–57.

Surani, M. A., Reik, W. and Allen, N. D. (1988) Transgenes as molecular probes for genomic imprinting. *Trends Genet.*, **4**, 59–62.

Surani, M. A. H., Barton, S. C. and Norris, M. L. (1984) Development of reconstituted mouse egg suggests imprinting of the genome during gametogenesis. *Nature*, **308**, 548–50.

Surani, M. A. H., Reik, W., Norris, M. L. and Barton, S. C. (1986) Influence of germ line modifications of homologous chromosomes on mouse development. *J. Embryol. Morphol.*, **97** (Suppl.), 127–36.

Swain, J. L., Stewart, T. A. and Leder, P. (1987) Parental legacy determines methylation and expression of an autosomal transgene; a molecular mechanism for parental imprinting. *Cell*, **50**, 719–27.

Swamy, B. G. L. (1942) Female gametophyte and embryogeny in *Cymbidium bicolor* Lindl. *Proc. Indian Acad. Sci., Sect. B*, **15**, 194–201.

Swamy, B. G. L. (1943) Gametogenesis and embryogeny of *Eulophia epidendraea* Fischer. *Proc. Natl. Inst. Sci., India, Part B*, **9**, 59–65.

Swamy, B. G. L. (1949) Embryological studies in the Orchidaceae. 2. Embryogeny. *Am. Midl. Nat.*, **41**, 202–32.

Swanson, C. P., Merz, T. and Young, W. J. (1981) *Cytogenetics*, 2nd edn, Prentice-Hall, Englewood Cliffs, NJ.

Taliaferro, C. M. and Bashaw, E. C. (1966) Inheritance and control of obligate apomixis in breeding bufflegrass, *Pennisetum ciliare*. *Crop Sci.*, **6**, 473–6.

Thomas, P. T. (1940) Reproductive versatility in *Rubus* II. The chromosome and development. *J. Genet.*, **40**, 119–28.

Thompson, J. N. Jr and Woodruff, R. C. (1978) Mutator genes – pacemakers of evolution. *Nature*, **274**, 317–21.

Tiffney, B. H. and Niklas, K. J. (1985) Clonal growth in land plants: a paleobotanical perspective, in *Population Biology and Evolution of Clonal Organisms* (eds J. B. C. Jackson, L. W. Buss and R. E. Cook) Yale University Press, New Haven, pp. 35–66.

Trivers, R. (1988) Sex differences in rates of recombination and sexual selection, in *The Evolution of Sex* (eds R. E. Michod and B. R. Levin) Sinauer, Sunderland, MA, pp. 270–86.

Tschermak-Woess, E. (1949) Diploides *Taraxacum vulgare* in Wien und Niederosterreich. *Osterr. Bot. Zeitschr.*, **96**, 56–63.

Turnau, E. and Karczewska, J. (1987) Size distribution in some Middle

Devonian dispersed spores and its bearing on the problem of the evolution of heterospory. *Rev. Palaeobot. Palynol.*, **52**, 403–16.

Turner, B. C. and Perkins, D. D. (1979) Spore-killer, a chromosomal factor in *Neurospora* that kills meiotic products not containing it. *Genetics*, **93**, 587–606.

Tyron, A. F. (1968) Comparisons of sexual and apogamous races in the fern genus *Pellaea*. *Rhodora*, **70**, 1–24.

Urbanska-Worytkiewicz, K. (1974) L'agamospermie, systeme de repro-duction important dans la speciation des Angiospermes. *Bull. Soc. Bot. Fr.*, **121**, 329–46.

Van Groenendael, J. and de Kroon, H. (eds) (1990) *Clonal Growth in Plants: Regulation and Function*, SPB Academic Publishing, The Hague, The Netherlands.

Van Oostrum, H., Sterk, A. A. and Wijsman, H. J. W. (1985) Genetic variation in agamospermous microspecies of *Taraxacum* sect. *Erythrosperma* and sect. *Obliqua*. *Heredity*, **55**, 223–8.

Van't Hof, J. (1965) Relationships between mitotic cell duration, S Period duration and the average rate of DNA synthesis in the root meristem cells of several plants. *Exp. Cell Res.*, **39**, 48–58.

Van't Hof, J. and Sparrow, A. H. (1963) A relationship between DNA content, nuclear volume, and minimum mitotic cycle time. *Proc. Natl. Acad. Sci. USA*, **49**, 897–902.

Vasek, F. C. (1968) The relationship of two ecologically marginal sympatric *Clarkia* populations. *Am. Nat.*, **102**, 25–40.

Vasil, V. (1959) Morphology and embryology of *Gnetum ula* Brongn. *Phytomorphology*, **9**, 167–215.

Ved Brat, S. V. (1966) Genetic systems in *Allium*. II. Sex differences in meiosis. *Chrom. Today*, **1**, 31–40.

Vijayaraghavan, M. M. and Shukla, A. K. (1977) Absence of callose around the microspore tetrad and poorly developed exine in *Pergularia daemia*. *Ann. Bot.*, **41**, 923–6.

Vijayaraghavan, M. R. and Prabhakar, K. (1984) The endosperm, in *Embryology of Angiosperms* (ed. B. M. Johri) Springer-Verlag, Berlin, pp. 319–76.

Vitt, D. H. (1968) Sex determination in mosses. *Michigan Bot.*, **7**, 195–202.

Wagner, W. H. (1974) Structure of spores in relation to fern phylogeny. *Ann. Mo. Bot. Gdn.*, **61**, 332–53.

Walker, T. G. (1966) Apomixis and vegetative reproduction in ferns. *Bot. Soc. Br. Isles, Conf. Rep.*, **9**, 152–61.

Walker, T. G. (1979) The cytogenetics of ferns, in *The Experimental*

Biology of Ferns (ed. A. F. Dyer) Academic Press, London, pp. 87–132.

Walters, M. S. (1950) Spontaneous breakage and reunion of meiotic chromosomes in the hybrid *Bromus trinii x B. maritimus*. *Genetics*, **35**, 11–37.

Warmke, H. E. (1954) Apomixis in *Panicum maximum*. *Am. J. Bot.*, **41**, 5–11.

Watson, E. V. (1971) *The Structure and Life of Bryophytes*, 3rd edn, Hutchinson, London.

Weimarck, G. (1973) Male meiosis in some amphimictic and apomictic *Hierochloe* (Gramineae). *Bot. Not.*, **126**, 7–36.

Weismann, A. (1889) Essays upon heredity and kindrid biological problems, translated by E. B. Poulton, S. Schonland and A. E. Shipley, Clarendon Press, Oxford.

Went, D. F. (1975) Blastoderm formation in artificially activated eggs of *Pimpla turionellae* (Hym.). *Devel. Biol.*, **45**, 183–6.

Went, D. F. (1982) Egg activation and parthenogenetic reproduction in insects. *Biol. Rev.*, **57**, 319–44.

Went, D. F. and Krause, G. (1973) Normal development of mechanically activated, unlaid eggs of an endoparasitic Hymenopteran. *Nature*, **244**, 454–5.

Went, D. F. and Krause, G. (1974) Egg activation in *Pimpla turionellae* (Hym.). *Naturwissenschaften*, **61**, 407–8.

White, M. J. D. (1938) A new and anomalous type of meiosis in a mantid, *Callimantis antillarum* Saussure. *Proc. R. Soc. Lond. [Biol]*, **125**, 517–23.

White, M. J. D. (1965) Chiasmate and achiasmate meiosis in African eumastacid grasshoppers. *Chromosoma*, **16**, 271–307.

White, M. J. D. (1970) Heterozygosity and genetic polymorphism in parthenogenetic animals, in *Essays in Evolution and Genetics in Honour of Theodosius Dobzhansky* (eds M. K. Hecht and W. C. Steere) North Holland, Amsterdam, pp. 237–62.

White, M. J. D. (1973) *Animal Cytology and Evolution*, 3rd Edn, Cambridge University Press, Cambridge.

White, M. J. D., Contreras, N., Cheney, J. and Webb, G. C. (1977) Cytogenetics of the parthenogenetic grasshopper *Warramaba* (formerly *Moraba*) *virgo* and its bisexual relatives. *Chromosoma*, **61**, 127–48.

White, R. A. (1979) Experimental investigations of fern sporophyte development. in *The Experimental Biology of Ferns* (ed. A. F. Dyer) Academic Press, London, pp. 505–49.

Whitehead, R. A. and Chapman, G. P. (1962) Twinning and haploidy in *Cocos nucifera* Linn. *Nature*, **195**, 1228–9.

Whittier, D. P. (1970) The rate of gametophyte maturation in sexual and apogamous species of ferns. *Phytomorphology*, **20**, 30–5.

Willemse, M. T. M. and de Boer-de Jeu, M. J. (1981) Megasporogenesis and early megagametogenesis. *Acta Soc. Bot. Pol.*, **50**, 105–14.

Willemse, M. T. M. and Van Went, J. L. (1984) The female gametophyte. in *Embryology of Angiosperms* (ed. B. M. Johri) Springer-Verlag, Berlin, pp. 445–74.

Williams, G. C. (1975) *Sex and Evolution*. Princeton University Press, Princeton.

Willson, M. F. and Burley, N. (1983) *Mate Choice in Plants*. Princeton University Press, Princeton.

Wilms, H. J. (1981) Ultrastructure of the developing embryo sac of spinach. *Acta Bot. Neerl.*, **30**, 75–99.

Winge, O. (1917) The chromosomes: their numbers and general importance. *C. R. Trav. Lab. Carlsberg*, **13**, 131–275.

Wolf, P. G., Sheffield, E. and Haufler, C. H. (1991) Estimates of gene flow, genetic substructure and population heterogeneity in bracken (*Pteridium aquilinum*). *Biol. J. Linn. Soc.*, **42**, 407–23.

Woodcock, C. L. F. and Bell, P. R. (1968a) Features of the ultrastructure of the female gametophyte of *Myosurus minimus*. *J. Ultrastr. Res.*, **22**, 546–63.

Woodcock, C. L. F. and Bell, P. R. (1968b) The distribution of deoxyribonucleic acid in the female gametophyte of *Myosurus minimus*. *Histochemie*, **12**, 289–301.

Woodruff, R. C. and Thompson, J. N. Jr (1980) Hybrid release of mutator activity and the genetic structure of natural populations. *Evol. Biol.*, **12**, 129–62.

Woodruff, R. C., Thompson, J. N. Jr and Lyman, R. F. (1979) Intraspecific hybridisation and the release of mutator activity. *Nature*, **278**, 277–9.

Wyatt, R. (1982) Population ecology of bryophytes. *J. Hattori Bot. Lab.*, **52**, 179–98.

Wyatt, R. (1985) Terminology for bryophyte sexuality: toward a unified system. *Taxon*, **34**, 420–5.

Wyatt, R. and Anderson, L. E. (1984) Breeding systems in bryophytes. in *The Experimental Biology of Bryophytes* (eds A. F. Dyer and J. G. Duckett) Academic Press, London, pp. 39–64.

Yudin, B. F. (1970) Capacity for parthenogenesis and effectiveness of selection on the basis of this character in diploid and autotetraploid maize. *Genetica*, **6**, 13–22.

Zander, R. H. (1984) Bryophyte sexual systems: -oicous versus -oecious. *Bryol. Beitre*, **3**, 46–51.

Taxonomic index

Subject index